SpringerBriefs in Earth Sciences

For further volumes:
http://www.springer.com/series/8897

Safiya M. Hassan

Sequence Stratigraphy of the Lower Miocene Moghra Formation in the Qattara Depression, North Western Desert, Egypt

 Springer

Safiya M. Hassan
Africa Earth Observatory Network (AEON)
Nelson Mandela Metropolitan University
Port Elizabeth
South Africa

and

Geology Department
Beni Suef University
Beni Suef
Egypt

Additional material to this book can be downloaded from http://extras.springer.com.

ISSN 2191-5369   ISSN 2191-5377   (electronic)
ISBN 978-3-319-00329-0   ISBN 978-3-319-00330-6   (eBook)
DOI 10.1007/978-3-319-00330-6
Springer Cham Heidelberg New York Dordrecht London

Library of Congress Control Number: 2013936517

Printed on acid-free paper

Springer is part of Springer Science+Business Media (www.springer.com)

For my Mother,
My second mother, Tahani Abdalla,

and

Professor Ahmed Abu Khadrah, Professor
Ronald Steel, Professor Nicholas
Christie-Blick,

and

The Moghra Team

Acknowledgments

As author, I want to express my profound sense of reverence to my supervisor and promoter, Prof. Dr. Ronald Steel, for his constant guidance, support, motivation, and untiring help during the course of my Ph.D. and my stay in Austin.

I am also deeply indebted to Prof. Dr. A. Abu Khadrah for his sustained supervision, for many useful discussions, and for reading the manuscript on Moghra Formation research. I am also very grateful to Prof. Dr. G. Abdel Gawad at the Geology Department, Beni Suef University, for his constant encouragement and support, extremely useful discussions, and suggestions on the manuscript, and to Prof. Dr. M. A. Hamdan for his great contributions in the fieldwork and his sincere supervision.

The author wishes to thank Dr. A. N. El Barkooky, for suggesting the research topic, close supervision, critical reviewing of the thesis, and fruitful discussions. Both Dr. El Barkooky and Prof. Dr. Hamdan have provided the author with insights into sedimentology and stratigraphy that will be useful throughout her career. They have also been open-minded and helpful advisors throughout the multiple stages of this research.

The author thanks Prof. Nicholas Christie-Blick, Department of Earth and Environmental Sciences, Lamont-Doherty Earth Observatory of Columbia University, for his guidance during the early stages of this thesis research. Steven Goldstein and Yue Cai are thanked for the laboratory support at Lamont-Doherty Earth Observatory of Columbia University.

The author is indebted to the generous help, insight, and useful discussions on the thesis by Prof. William Fisher, Prof. Charles Kerans, Prof. Robert Folk, Prof. Earle McBride, Prof. Mark Helper, and Prof. Christopher J. Bell. In addition, I thank my colleagues at Jackson School of Geosciences, University of Texas at Austin; Cornel Olariu, Shaikh Muhammad, Andy Petter, Nancy Elder, the head librarian of the life science library, and the rest of the Jackson School team.

I would like to thank my mother for my continued encouragement and support. I would also like to thank Dr. Nahla Shallaly, Mohamed Abdel Gouad (Cairo University), and my godfather Said Mehaseb and his wife Fouzia Benzekri for their continuous help. And I would also like to express my profound appreciation to Tahani Abdalla for her special encouragement.

Finally, the author would like to thank the present host, Prof. Maarten J. de Wit at Africa Earth Observatory Network (AEON) and Faculty of Science, Nelson Mandela Metropolitan University, for his encouragement and support to publish this manuscript. And I would also like to thank my Colleague Callum Anderson for his support and friendship, Geosciences at Nelson Mandela Metropolitan University.

For financial support I want to acknowledge the National Science Foundation (NSF) the Egyptian Mission Department, RioMAR Industry Consortium at the University of Texas at Austin, and the Egyptian Ministry of Higher Education and Research.

Supervision Committee

Prof. Dr. Ahmed Mokhtar Osman Abu Khadrah.

Geology Department, Faculty of Science, Cairo Uinversity.

Prof. Dr Gouda I. Abdel Gawad.

Head of Geology Department, Faculty of Science, Beni-Suef Uinversity.

Prof. Mohamed A. Hamdan.

Geology Department, Faculty of Science, Cairo University.

Dr. Ahmed Niazy El-Barkooky.

Geology Department, Faculty of Science, Cairo Uinversity.

Prof. Dr. Ronald Steel.

Department of Geolgical Sciences, Jackson school of Geosciences, The Uinversity of Texas at Austin.

Contents

Chapter 1
Introduction

1.1 Introduction

The northern cliffs of the Qattara Depression exhibit excellent outcrops of the Lower Miocene Moghra Formation, which is known for its fossil vertebrates. Despite this successful focus on the vertebrate of Moghra area, there is still a noticeable gap in our knowledge about Lower Miocene sedimentology and sequence stratigraphy of this area. The literature about the sedimentology and sequence stratigraphy of the Moghra Formation has been sparse, despite some excellent work over the years by academic and petroleum workers. Moreover, the studied area is within what was a front-line of World War II, where mine fields and war relics are scattered and cover wide reaches. This has resulted in limited geologic mapping in the past. Thus, great attention is paid in this study to establish a robust sedimentology and high-resolution sequence stratigraphic framework for the Lower Miocene Moghra Formation. Included are works based on outcrops and, most importantly, new sedimentological and chronostratigraphic information not previously available. Moreover, the palaeogeographic reconstruction and facies distribution of the Lower Miocene in this area (onshore Mediterranean) will directly impact the offshore petroleum exploration strategies in terms of reservoir prediction and proximal–distal facies variation.

1.2 Location

The area under study, covering about 40 km², is limited by Latitudes 30° 10′ and 30° 30′ N and Longitudes 28° 30′ and 29° E (Fig. 1.1).

The area is accessible from El Hammam and Alamein (about 60 km to the south) on the Mediterranean coastal zone, by unpaved tracks, traversing the studied area. Moreover, it can be reached from Wadi El Natrun, to the east, by a

S. M. Hassan, *Sequence Stratigraphy of the Lower Miocene Moghra Formation in the Qattara Depression, North Western Desert, Egypt*, SpringerBriefs in Earth Sciences, DOI: 10.1007/978-3-319-00330-6_1, © The Author(s) 2013

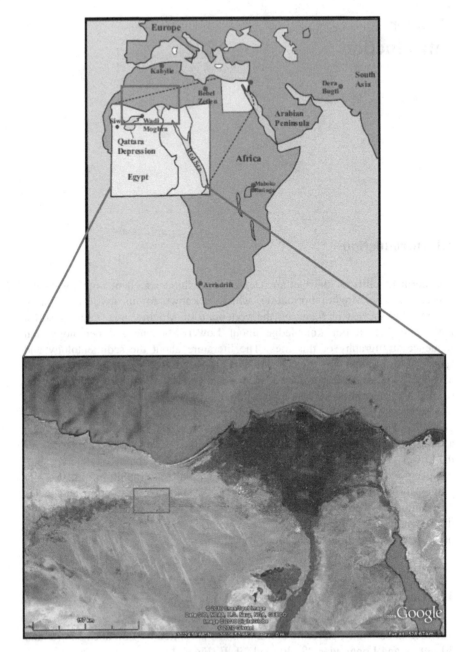

Fig. 1.1 Location map of the study area (*Red square*, modified after Sanders and Miller 2002)

poor and undefined track. Now, the area is easily accessible after some urban development has moved towards the area, and the low-lying land of the area has been cultivated for different type of crops.

1.3 Aim and Scope of Study

Despite the progress made on understanding the paleontology and general stratigraphy of Moghra, a number of research questions remain to be addressed regarding the depositional facies and their time/space distribution. No sequence stratigraphic studies have previously been undertaken in the Moghra area.

The aim of the study is to establish a robust high-resolution sequence stratigraphic framework for the Lower Miocene Moghra Formation and to reconstruct the palaeogeography for that time. Moreover, the palaeogeographic reconstruction and facies distribution of the Lower Miocene in that area (onshore Mediterranean) will directly impact the offshore petroleum exploration in terms of reservoir prediction and proximal distal facies variation.

The present work provides a large amount of new data on all aspects of the Moghra Formation, ranging from stratigraphy to chemostratigraphy to sequence stratigraphy. Perhaps one of the important contributions is the isotope stratigraphy. There is a lack of suitable material for absolute dating of the Moghra Formation. Nevertheless, indirect dating using strontium isotope stratigraphy has been possible. It has been known, particularly for the mid-Cenozoic, that the strontium isotope sea-water curve changes rapidly with time, and so is particularly appropriate for geochronology. In this time interval, resolution of stage boundaries is better than 0.5 m.y. (Howarth and Mcarthur 1997; Oslick et al. 1994), making the method a powerful tool for improving correlation between chronostratigraphic time-scales and biostratigraphy. So combined biostratigraphic and strontium isotope analysis has resulted in a much firmer age setting for the Moghra Formation.

1.4 Methodology

Different approaches will be followed in reaching the targets of the present study through field work, work flow of sequence stratigraphy and laboratory analyses.

1.4.1 Work Flow

Sequence stratigraphic principles can be readily applied to outcrops by using the following workflow steps.

1. Facies analysis. In order to correctly identify the depositional environments, a facies model needs to be constructed, so that proximal, distal and lateral facies relationships can be assigned. A facies model is constructed by careful study of all elements of the sedimentary succession, identifying the signals of physical processes that were taking place during the deposition, and interpreting environmental systems. In a vertical succession, using the facies model, a shallowing- and or deepening upwards succession can be defined.

2. Stacking patterns, parasequences and parasequences sets can be recognized and recorded.
3. Following the documentation of parasequences and the processes that produced them, systems tracts can be identified based on the overall stacking patterns of the parasequence sets, and how the stacking pattern changes through the succession.
4. Surfaces of maximum regression (sometimes equivalent to the sequence boundaries, but not where there are lowstand deposits) and surfaces of maximum transgressions (maximum flooding surfaces) can be defined and supported by field observations. The succession then can be divided into sequences.
5. Accurate delineation of the age and timing of subaerial exposure, unconformities and truncation surfaces is a powerful tool for detecting sea-level change and sediment supply change.
6. Different time scales are defined as orders and used in the linkage between various scales of cyclo-stratigraphy. Special emphasis is given to the relationship between third-order (1–10 m.y. period) depositional cycles and their component fourth-order (0.1–1 m.y.) cycles through detailed stratigraphic analyses of Moghra Formation.
7. Building the concept of an early Miocene integrated depositional and sequence stratigraphic model in the study area.
8. Finally, The sequence stratigraphic framework is developed, and then the results are compared with the global sea level curves of Haq et al. (1987, 1988) and Pekar and Christie-Blick (2008).

1.4.2 Field Work

Field work was carried out through the following steps in order to complete the study of the Lower Miocene of Moghra Formation.

1.4.2.1 Reconnaissance within the Study Area

This has been carried out using a map produced by the satellite image of Lower Miocene Moghra Formation in The Qattara Depression, North Western Desert to select suitable section localities, as well as to check the previous work. In addition, we used the topographic map El Moghra, sheet NH35: L1. Scale 1:100,000.

1.4.2.2 Measurement and Sampling of the Selected Sections Through the Following Steps

a. Stratigraphic sections have been measured, and described across all the studied area from the East to the West.

b. More than 500 rock samples have been collected for different litho facies. They show that there is remarkable facies change within every section.
c. Palaeocurrent measurements are also used for the palaeogeographic interpretations.
d. The best sections in the studied area were selected and a detailed sedimentological and palaeontological study was conducted in order to provide a serious base for environmental reconstruction.
e. The depositional environment has been modeled in relation to the sequence stratigraphic framework.

1.4.3 Laboratory Work

These were carried out for some special samples of certain interest to fine-tune the characterization of the units. The results are used for supporting and enhancing stratigraphic and sedimentologic interpretations. Among laboratory analyses used are:

1. The collected hand specimens and polished slabs have been examined by eye to detect the variations in their lithology, colours, sedimentary structures and textures (at mesoscopic scale).
2. Petrographic microscopy of thin section. Most samples were subjected to microscopic examination in thin sections. The results obtained are used to determine the petrography of the rocks and to shed more light on the framework of sedimentation and history of diagensis.
3. Chemostratigraphy dating of some samples of good preservation.
4. We have determined the $^{87}Sr/^{86}Sr$ ratios and derived absolute ages for more than 15 macrofossil samples collected from several biostratigraphically dated Lower Miocene sections in the Moghra area. Measurement on a mass spectrometer of the strontium (Sr) isotope ratio $^{87}Sr/^{86}Sr$ in marine fossil shells has provided absolute ages for those shells.
5. Paleontology investigation of selected samples for more accurate age determination.

1.4.4 Previous Works

The Miocene sequence in the Western Desert has been studied by many previous workers: Fourtau (1920), Shata (1955), Said (1960, 1961a, b, 1962a, b, 1971 and 1990), Abdallah (1996), Marzouk (1970), Philip et al. (1973), Philip and Darwishe (1973), Omara and Sanad (1975), El-Khashab (1977), Misak (1979), Al-Wakeel (1989), Abu-Zeid and Al-Wakeel (1992), Abdallah (1996), Sharaf (1995), Miller and Simons (1996), El-khoriby (2004), Abd-Alla (2001).

Fourtau (1918), published the first treatise on the Moghra vertebrate fauna (see also Blanckenhorn, 1901), and in 1920 he re-published his previous account with additional appendices.

Shata (1955) discriminated the Miocene rocks of the Western Desert into two horizons: 1. An upper horizon of Middle Miocene shallow marine limestone of maximum development of 300 m thick, becoming partially fluviatile east of longitude 29° E.

2. A lower horizon of Lower Miocene mainly composed of shallow marine fossiliferous limestone and marl changing to sand and silt south and east of the Qattara Depression. It attains a thickness of approximately 300 m.

Said (1962a, b) divided the Miocene sediments of the Western Desert of Egypt into two distinct rock units from base to top: the Moghra Formation and Marmarica Limestone Formation (Table 1.1). Moghra Formation covers all the facies exhibited by the Lower Miocene sediments of the North Western Desert. He distinguished an "estuarine", marine, reefal and open bay facies. In addition, a genuine fluviatile facies, given the name Gebel Khashab Redbeds, is also recognized. Furthermore, he concluded that the lowermost Lower Miocene (Aquitanian) is not found. The Moghra Formation is of Burdigalian age. The Marmarica limestone unit which caps the underlying Moghra clastics shows little lithological change throughout its entire thickness and width which covers almost 6° of longitude. It changes from solid limestone unit in the west to a sandy limestone unit in the east. The lower clastic Moghra Formation shows areal variations with regard to its lithology and its faunas. This variation is intimately connected with the geological history of the region during lower Miocene time. Two facies can be recognized: a proper marine facies in the west and a fluvio-marine facies in the east.

Abdallah (1966) stated that the Lower Miocene Moghra Formation is deposited under fluviomarine, semicontinental and estuarine conditions. This is proved by the presence of vertebrate remains, marine macro and micro-fossils, wood remain, thin gypsum seams or lentils and thin carbonaceous and brown coal very thin seams and specks. The Moghra Formation is Burdigalian Age and also the lowermost Lower Miocene (Aquatanian) is not found. From the tectonic viewpoints, no major true structural displacements (faulting or folding) are recognized. Nevertheless, few exceptions are observed, represented by fracturing and jointing caused by epeirogenic movement and/or mild post-Miocene extensional forces (Pliestocene?). Jointing (fractures in general sense), landslides and minor pseudofolding (i.e. upwards gentle doming "Aufwolbung") are the main surface structural features recorded.

Norton (1967) delineated the following three rock units (from base to top):

1.The Moghra group (Oligocene to Early Miocene) is represented by interbedded sandstone, marls, shale and gypsum, with occasional limestone beds and several fossil wood and vertebrate remains. Oligocene basalt is interbedded in the Moghra group west of Giza where it conformably overlies Upper Eocene deposits.

2. The Qaret Shushan Formation (Oligocene to Early Miocene) conformably overlies the marine Oligocene shale and is represented by interbedded shale,

Table 1.1 Classification of the miocene deposits in the Western Desert. (After Al-Wakeel 1989)

Said (1962a)		Marzouk (1970)		Omara & Ouda (1972)				Misak (1979)			
Age	Formation	Age	Formation	Epoch	Age	Stage	Formation	Age	Formation	Member	
Middle Miocene	Marmarica Limestone	Middle Miocene	Marmarica Limestone	Miocene	Late Miocene	Touronian / Messinian	Al-Jaghbub Formation	Middle Miocene	Al-Jaghbub Formation	Upper. L. S. Mm	Deir El Abyiad member
					Middle Miocene	Langhian				Basal L.S. Mm	Ras El Husan member
Early Miocene	Moghra Formation	Early Miocene	Mamura Formation		Early Miocene	Burdigalian	Mamura Formation	Early Miocene	Moghra Formation	Upper. S. S. Mm	Qaret El Rikab member
			Abu-Subeiha Formation			Aquitanian	Khalda Formation			Clay Mm	Deir El Tarfaya member
			Moghra Formation	Oligocene	Late Oligocene		Shushan Formation			Basal. S. S. Mm	El Raml member
			Abu-Sennan Formation								

sandstone and limestone of open marine facies to the north and northwest, reefal to the west and lagoonal to estuarine towards the south and east.

3. The Giarabub Formation (Early to Middle Miocene) is represented by fossiliferous calcareous with interbedded clays, marl and calcareous sandstone; all are of shallow neritic to marginal environment. The Formation conformably overlies the Qaret Shushan Formation and unconformably underlies the Pleistocene sediments.

Marzouk (1970) classified the Miocene rocks of the Western Desert into the following rock units (Table 1.1): (1) Marmarica Formation (2) Mamura Formation (3) Abu Subeiha Formation (4) Moghra Formation and (5) Abu Sennan Formation. The Marmarica Formation conformably overlies Mamura Formation and it is exposed on the surface. The Marmarica Formation is shallow neritic to marginal marine. It is Middle to Lower Miocene. Mamura, Abu Subeih and Abu Sennan formations were a new formational names proposed by Marzouk (1970) to represent the Lower Miocene marine facies, mixed lithology of fluvio-marine to shallow marine facies Lower Miocene, with sand/shale ratio of not more than 1/3 and very rare or no limestones and continental-mainly eolian facies of both the Lower and Middle Miocene in the Western Desert (Abu Sennan Formation). They conformably overlie Alam El-Bueib Formation (Oligocene-Upper Eocene) and conformably underlie Marmarica Formation (Middle Miocene). Mamura Formation detected from the subsurface except very few surface sections which show its topmost parts. Marzouk (1970) redefined the Moghra Formation of Said (1961a) and restricted it to represent the fluvio-marine facies only of the Miocene in the Western Desert.

According to El Bassyony (1971), the Tertiary rock units represent deposition in a very large-scale, regressive succession as displayed by the shallow marine littoral and restricted lagoonal facies of the Eocene sediments, the channel facies of the Oligocene and the continental sediments of the Lower Miocene, the Miocene deposits are represented by the Moghra Formation of Lower Miocene age, and these are found to be formed as a slightly undulating to nearly flat surface covered by thin sheets of dark brown and light yellow flint and quartz pebbles and gravels.

Omara and Ouda (1972) subdivided the Miocene sequence into the following three formation from base to top (Table 1.1):

1. The Khalda Formation (Aquitaninan) is composed essentially of clays and sandstones containing shallow marine planktonics (e.g. *Globigerinoides Pimordius* and *G. altiapertura*) and benthoincs (e.g. *Elphidium advenum, Cibicdes boueanus, Rotalia andouini* and *Elphidium antonium*). The unit grades into fluviatile facies eastwards through transitional mixed facies. This formation is encountered west of longitude 29° E overlying and underlying conformably the Shushan Formation and Mamura Formation respectively.

2. The Mamura Formation (Burdigalian to Langhian) is made up of alternation of grey greenish shale and greenish limestones of shallow marine origin as indicated by the planktonics (e.g.) *Globigerinoides altiapertura, G. trilobus, G. sicanus* and *Miogypsina globulina*) and benthonics (e.g. *Rotalia andonini, Elphidium antonium*). It grades eastwards into a fluvio-marine facies. In the central and

eastern portions of the north Western Desert, the uppermost part of the Mamura Formation reflects deposition in a rather shallow water environment and yields a rich benthonic assemblage of a Langhian age. The formation underlies conformably the Al-Jaghbub Formation. A part of this formation is stratigraphically equivalent to that of Marzouk (1970).

3. The Al-Jaghbub Formation (Langhian-Helvetian): Omara and Ouda (1972) designated the Al-Jaghbub series of Desio (1928) in Lybia to its equivalent in the Western Desert that was named by Said (1962a) as the Marmarica Formation. It becomes partly dolomitic and it increases to a great thickness westward where it attains about 600 ft. On the contrary to the east there is a marked thinning and a gradual increase in its sand-content; near the Moghra Oasis it is formed of 17 ft of thick sandy limestone. It replaces "Marmarica limestone Formation and its type locality is the high escarpment bordering Al Jaghbub Oasis, which is located very close to the Egyptian-Libyan border. It is conformably overlain by white sandy limestone of Late Miocene "Tortonian – Messinian" age.

Furthermore, Omara and Ouda (1972) restricted the Shoushan Formation to the solid, white, reefal limestone, capping the marine Upper Eocene–Oligocene shales "Ghazalat Formation", and underlying the Aquitanian clays and sandstones "Khalda Formation". It yields a shallow-water benthonic association of reefal nature, marking, therefore, a great shallowing of the sea at the end of the Oligocene. In the eastern portion of the northern Western Desert, it changes to sandy shales, due to the prevalence of fluviatile conditions.

The Stratigraphic Subcommittee of the National committe Geological Sciences of Egypt (1974) used the following classification for the Miocene succession in the Western Desert (from base to top): (1) The Moghra Formation (Early Miocene) is restricted to the fluvio-marine facies in the Western Desert. (2) The Mamura Formation that was established by Marzouk (1970) to represent the shallow marine to neritic facies in the Western Desert and is laterally equivalent to the Moghra Formation. (3) The Marmarica Formation (Middle Miocene) that was established by Said (1962a) to represent shallow neritic to marginal marine limestones and marls with a number of oyster banks and rich in neritic and reefal assemblages of invertebrates.

Salem (1976) stated that the Miocene rocks cover many outcrops of the Western Desert. The lower Miocene is represented by the sandy Moghra Formation (which is of Oligocene age in part) and is best exposed in the cliffs of the Qattara Depression. The middle Miocene is predominantly sequences from carbonate rock and is referred to as the Marmarica Limestone (see Said 1961b; Said 1962a). The Marmarica Limestone, which covers vast expanses of the Western Desert, slopes gently toward the Mediterranean Sea. The lower Miocene sedimentary rocks in the subsurface of the Western Desert are made up of sandstone and shale with thin intercalations of limestone and the lower part of the section is more shaley. The lower Miocene strata are about 615 m thick in well WD-5. These sedimentary deposits are thin westwards and are replaced partly by carbonate beds. East of WD-5, near the western flank of the Nile delta, lower Miocene strata is exclusively sandstone and shale; thickness of the section has decreased to about 307 m, e.g.,

Wadi El Natrun well. In contrast, the middle Miocene is represented by a limestone section, the Marmarica Limestone, which thickens westwards.

Trying to reconstruct the Oligocene to Early Miocene paleogeography, Salem (1976) mentioned that the inherited northwest- southeast structural trend became well developed and affected the sedimentation. A river system flowing northwest debouched deltaic sediments into the central part of the Western Desert, which then was an embayment of the Mediterranean during the Oligocene. This embayment had irregular bottom topography, which can be termed immature shelf. Longshore currents from west to east helped to redistribute the sediments in the Oligocene embayment. The same current pattern also prevailed during Miocene. The early Miocene deltation continued in a pattern similar to that of the Oligocene, but with gradual shift in sedimentation east and northeast. Concurrently, a carbonate platform developed toward the west. In addition, in the middle Miocene, a sudden shift in the river course occurred, presumably as a result of faulting. This sudden shift initiated deposition of the middle Miocene Nile delta near the site of its modern counterpart. The abandoned deltaic and parahc sediments on the west were onlapped by carbonate deposits that advanced from west to east.

Misak (1979) subdivided the Moghra Formation into the following three members from base to the top as follow (Table 1.1):

1. The basal sand member (El-Raml member). This member consists of variegated sands and sandstone (215 m thick) with clay interbeds and minor limestone bands. The sandstone is occasionally argillaceous, glauconitic and/or ferruginous.

2. The clay member (Deir El-Tarfaya member) is made up of massive sandy clays (38 m thick) which are micaceous and gypseous.

3. The upper sand member (Qaret El–Rikab member) consists of cross-bedded friable sandstone of loose quartz sand (75 m thick), mostly of yellow colour, intercalated with thin streaks of grey clay, ferruginous and calcareous sandstone.

Misak (1979) also subdivided the Mamura Formation of Marzouk (1970) into three non-nominated units. The basal unit (25 m thick) is made up of grayish clay and sandy limestone, the middle unit (25 m thick) is composed of grayish carbonaceous clays and the upper unit (9 m thick) consists of grayish clayey sandy limestones.

Khalifa and Abu Zeid (1985) illustrated the environmental models of the Lower and Middle Miocene sediments in the area northwest of Wadi El Natrun. Two models have been constructed for the Lower Miocene Raml/Moghra Formations and the Middle Miocene Marmarica/Hagif Formations. The Early Miocene model consists of two equivalent environments, fluvial and mixed fluviomarine. The fluvial environment prevailed southwards depositing Raml Formation. The facies is sandstones with conglomerate and claystone intercalations enriched in silicified wood and vertebrate remains. The northern equivalent environment is the mixed fluviomarine of the Moghra Formation. In such an environment, there were alternating phases between the marine fossiliferous clays and the fluvial sands. The Middle Miocene model involves three laterally equivalent environments. These are coastal marine, shelf lagoon and open inner shelf. They extend from "shallower" in the east to "deeper" in the west. The coastal marine and shelf lagoon are attributed to Hagif Formation. The open inner shelf ascribed for the Marmarica Formation which extends west of longitude 29 E. This environment seems to be deeper than those of Hagif Formation, marking the deepening of the shelf towards the west.

Al-wakeel (1989) studied three successions of the Moghra Formation (Early Miocene) overlain by a considerable thickness of the Marmarica Formation (Middle Miocene). The first succession (208 m) is well developed at the eastern promontory of the Minquar Abu-Duweis, which is a part of the northern wall of the Qattara Depression. Al-wakeel (1989) recommended this succession as a new stratotype for the Moghra Formation. The second and third successions are the southwest Minquar Abu-Duweis section and the Qaret El-Hemeimat section in order to test the lateral variations in the lithofacies characteristics of the sediment. The Moghra Formation lithologically is a thick, predominantly clastic rock unit consisting of sands and sandstones intercalated with siltstones. He delineated eight facies from the vertical lithologic variations in the Moghra Formation. The eight facies from the base to top as follows: (i) the lower arenaceous unit; (ii) the lower calcareous unit; (iii) the argillaceous arenaceous unit; (iv) the middle calcareous unit; (v) the lower argillaceous unit; (vi) the upper calcareous unit; (vii) the upper argillaceous unit; and (viii) the upper arenaceous unit. The lower arenaceous unit (~ 35 m thick) consists of sands and sandstones containing ferruginated conglomeratic bands. The sandstone is almost devoid of marine fauna but contains vertebrate bone fragments and silicified tree trunks which lie horizontally trending N70 W. Small scale cross stratification is recorded in the lower part and large scale cross stratification is recorded in the upper unit. The lower calcareous unit (~ 17 m thick) is made up of calcareous, highly fossiliferous sandstones and thin limestone interbeds. The argillaceous-arenaceous unit (~ 74 m thick) is composed of claystones alternating with cross bedded, occasionally calcareous sandstones encompassing three slightly conglomeratic bands. Tabular and trough cross bedding, climbing ripples and deformation structures are encountered. The middle calcareous unit (~ 16 m thick) is made up of sandy limestones which laterally changes in the southwest into highly calcareous mudstones. The lower argillaceous unit (~ 22.5 m thick) consists of ferruginous, gypsiferous siltstones which become sandy and gravelly in the upper unit part of the unit that contains also vertebrate bone fragments. The upper calcareous unit (~ 7 m thick) is made up of highly calcareous, fossiliferous siltstones. The upper argillaceous unit (~ 17 m thick) is composed of calcareous, gypseous claystones alternating with thin sand streaks. The upper arenaceous unit (~ 19.5 m thick) is made up of calcareous, glauconitic sandstones containing a conglomeratic band and interbedded with gypseous claystone layers. The Marmarica Formation attains a thickness of about 53 m in the East Minquar Abu-Duweis section while in the Qaret El-Hemeimat section it is represented by a 27 m thick succession. The rock unit is composed of hard, argillaceous, sandy and dolomitic limestones. Layers of marls, claystones and sandstones are encountered in the lower part of the formation and eastwards in the Qaret El-Hemeimat sequence.

Albritton et al. (1990) mentioned that the Moghra Formation is predominantly a clastic unit consisting of fine- to medium grained, reddish-brown, usually friable sandstone and siltstone. Many of the units are cross-bedded, and some contain numerous small- to medium-sized, well-rounded quartz and chert pebbles. Large accumulations of petrified logs, 20–75 cm in diameter and as much as 10 m in length, are present in several stratigraphic horizons. The trunks lie nearly parallel to each other, with a general east-northeast orientation. Fossil vertebrates are present in the Moghra Formation, especially in the upper part. Large animals, such as hippopotamus, are present together with micro-vertebrate remains, including rodents. Some layers of carbonate rocks within the upper Moghra Formation (fluviatile and fluvio-marine facies) contain marine and fluvio-marine fossils.

Said (1990) noted that the maximum marine transgression of the Miocene epoch occurred during the Burdigalian when the sea covered large areas of north Egypt and overflowed into the newly formed Gulf of Suez. In addition, a large part of the area was under the influence of fluvial systems that formed the landward edge of a wave-dominated delta plain covering the eastern part of the north Western Desert. Said mentioned that the earliest Miocene sediments of Aquitanian age are of limited areal distribution. They are recorded with certainty in the north Delta embayment. Aquitanian strata are also recorded in one locality (Gebel Homeira) in the Cairo-Suez district (Sadek 1968). Here they assume a reefal carbonate facies and carry the characteristic fossil *Miogypsina tani*.

Said (1990) stated that the North Western Desert: Basin developed to the south of the Mediterranean coastal high, an old marginal offset which was active during the Paleogene. During the early Miocene, clastic sedimentation prevailed. A change of the climate and a reactivation of the coastal high during the middle Miocene left the north Western Desert a distinctive basin in which clastics were deflected and organogenic carbonate deposits accumulated.

Rizk and Davis (1991) stated that the Moghra formation occupies most of the floor of the Qattara Depression, extending to the east and dipping beneath younger formation to the north. Moghra Formation boundaries are well-known because of oil expolartion in northwestern Egypt and are described by Joint-Venture Qattara (1981) as follows:

1. The top of the Moghra Formation is mostly exposed except in the area to the north of Qattara depression, where it is overlain by the Middle Miocene limestone of the Marmarica Formation.
2. The base of the Moghra formation is marked by Oligocene shale of the Dabaa Formation. The thickness of the Moghra Formation varies according to the buried structure on which it was deposited.
3. To the north, the Moghra Formation grades sharply into a less permeable, clayey facies, expecially along the Mediterranean Sea coast, and is overlain by cavernous limestone of the Middle Miocene Marmarica Formation.
4. To the south and southwest, the Moghra Formation is limited by the progressively top of the Oligocene shale of the Dabaa Formation. Oligocene basalt

flows and fault escarpment associated with the Bahariya-Abu Roash uplift from the southeastern boundary of the Moghra Formation.

5. To the east, the thickness of moghra Formation decreases gradually and its clastic nature changes into a less permeable, clayey facies.

To the west, the Moghra Formation interfingers with a less permeable limestone and shale sequence. The top of the underlying shale of the Dabaa Formation is higher westward, delineating the western boundary of the Moghra Formation.

Hafez (1992) studied the Early and Middle Miocene sequence exposed at Qaret El-Kish, North Qattara Depression, Western Desert. From his study, he concluded that the Moghra Formation of Early Miocene age (250 ms thick) is a thick unit of clastics (sandstones intercalated with clay, limestone and gypsum) that underlie the Marmarica Formation of Middle Miocene age (62 ms thick) and made up of carbonate facies. The Moghra Formation of the Early Miocene succession consists of Quartz arenite, litharenite and quartz wackes, sandy sparite and calcareous sandstone. On the other hand, the Marmarica Formation is composed of sandy biosparite, sandy algal biomicite, sandy dolomitized biomicrite and sandy microsparite. The light fractions are composed of quartz (91.4 and 98.6 %) together with subordinate amounts of feldspars (1.6–6.7 %) while the heavy mineral assemblages recorded in the studied Miocene sands include (arranged a decreasing in order of abundance), opaques, zircon, tourmaline, rutile, staurolite, garnet, kyanite and hornblende. He believed that the Moghra Formation is of fluviatile origin especially in its lower part. It was interrupted by shallow marine sedimentation as a result of limited transgressions of the Early Miocene sea causing the deposition of different calcareous units with some marine fauna. On the other hand the upper parts of the Moghra Formation were deposited in a shallow marine environment.

Miller and Simons (1996) stated that in the early Miocene, however, new primates, carnivores, artiodactyls and perissodactyls appeared in the African record for the first time. In fact, much of the present-day African fauna, and many of the key evolutionary and migratory events that have shaped that fauna, can be traced to the early Miocene. Some of these events include: the origin and early evolution of the Old World monkeys, a Miocene radiation of large and small-bodied apes. Well known vertebrate taxa recovered from Moghra thus far include: anthracotheres, rhinocerotid perissodactyls, proboscideans, the ceropithecoid *Prohylobates tandyi* and an as-yet-unnamed ape. Recent work at Moghra has enhanced the known fauna with speciments representing suids, viverrids, amphicyonids, tragulids, sivatheriids and palaeomerycids. Geologically, some of the sediments at Moghra are likely to represent low-energy fluviatile, estuarine and lagoonal conditions, as judged by the presence of sandstones, siltstones and calcareous shales (Hantar 1990). This depositional interpretation is supported by the mix of primarily marine animals such as sharks and rays, along with largely freshwater animals such as catfish and crocodiles, as well as a continental mammalian fauna. Also, many of the common aquatic vertebrates recovered from Moghra, such as catfish and crocodiles, can tolerate slow or stagnant water with high sediment content, lending credence to the idea that fluviatile conditions at Moghra were likely to have been slow-moving. The large mammal fauna is dominated by a variety of anthracotheres, a group of artiodactyls that apparently preferred an aquatic or aquatic margin environment as they are found most commonly in fluviatile and lacustrine deposits (Pickford 1991). In addition, the abundance of silicified tree trunks at Moghra suggests that this fluviatile estuarine-lagoonal system was probably forested.

Abdeldayem (1996) concluded that: (1) The Miocene sedimentary rocks from the northeastern tip of the Qattara Depression carry a weak remnant magnetization that is carried mainly by goethite, haematite and titanomagnetite. (2) The isolated characteristic remnant magnetization might reflect the age of these rocks and seems to be in a good agreement with coeval directions from the stable African craton. (3) The obtained results suggest that, since the Early Miocene, tectonics have either played no significant role in the formation of the depression or resulted in movements that could not be detected palaeomagnetically. Preferably, weathering processes seem to have been strong and could have played the principal role in the development of the depression.

Abd-Alla (2001) studied the sedimentology, mineralogy and depositional environments of the Moghra Fornation at Minquar Abu Duweis and said that the Moghra formation is composed mainly of sandstones interbedded with shales and minor limestones. The main sandstone types are calcareous, ferruginous and fossiliferous dolomitic quartz arenites. Oolitic packstone, bryozoan boundstone and sandy fossiliferous dolostone microfacies are identified within the limestone facies. Concerning the heavy minerals, the Moghra Formation sediments may be considered as a mixture of two components, mature debris rich in ultrastable minerals (RTZ), which were derived by recycling from the pre-Miocene sediments and have been mixed with immature debris rich in epidotes derived from basement rocks in the south and from the Red Sea highlands. The diversity of source rocks argues for several channels of a large river system which existed during the early Miocene and may have introduced considerable amounts of clastics into the central part of the north Western Desert. Kaolinite is the predominant clay mineral found in the Moghra sediments. According to Abd-Alla (2001), the encountered kaolinite is of sedimentary origin (i.e. formed through weathering, transport and deposition). The fact that the kaolinite in the Moghra clastics is perfectly to very well crystallized suggests that the reported clay mineral is partly authigenic and supports its formation in continental or non-marine environment. However, the presence of montmorilnite in the Moghra limestone may point to formation in a marine environment. Finally, he (ibid author) suggested a fluvio-marine environment for the Moghra Formation where the Moghra clastics were deposited in a fluvial environment which was interrupted by a few brief marine invasions during the deposition of the two Moghra fossiliferous carbonate units.

1.5 Depositional Framework of Tertiary Basins (Miocene), Northern Egypt

The lower Miocene sequence in the Western Desert is very sandy as shown on the palaeogeographic map of Salem (1976) (Fig. 1.2). Oligocene sedimentation apparently had built a shelf with a more even bottom topography, a mature shelf. On this mature, early Miocene shelf, sediments possibly were distributed as broad, flat, thin, high destructive deltas (of Fisher et al. 1969; Scruton 1960) that were dominated by waves and long shore currents. Repeated fluvial deposition advanced across low-lying delta plain areas and reoccupied older fluvial sites (Figs. 1.2, 1.3, 1.4).

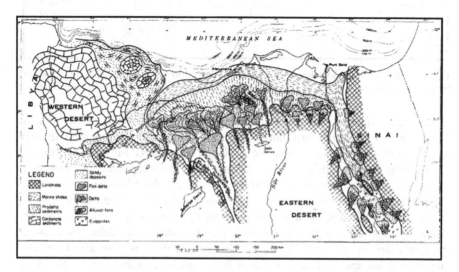

Fig. 1.2 Tentative depositional model for Early Miocene time, northern Egypt (modified from Salem 1976). AAPG©[1976], "reprinted by permission of the AAPG whose permission is required for further use"

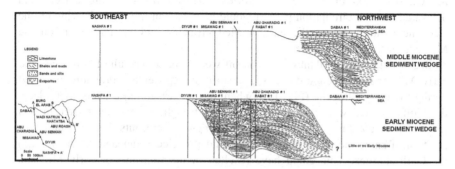

Fig. 1.3 Cross sections along line A–A′ (modified from Salem 1976). AAPG©[1976], "reprinted by permission of the AAPG whose permission is required for further use"

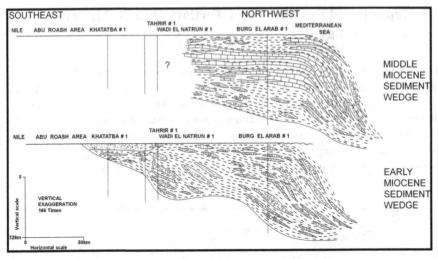

Fig. 1.4 Cross sections along line B–B′ (modified from Salem 1976). AAPG©[1976], "reprinted by permission of the AAPG whose permission is required for further use"

Carbonate material was being deposited in the northwest part of the Western Desert away from the main influence of terrigenous clastic sediment. The early Miocene setting is similar to the present relation between the Mississippi delta and the flanking Florida carbonate bank. Indeed, there are many similarities between the depositional schemes in the Gulf of Mexico and the Miocene and even the recent Mediterranean basin. One such similarity is the long shore-current system. In the Gulf of Mexico, the long shore current is from east to west; i.e., currents move from carbonate banks of Florida toward the Mississippi delta. This current transports deltaic sediments westward. At present, there is a similar longshore current that sweeps the coasts of Egypt from west to east and transports deltaic sediments of the Nile toward Port Said and farther east. The circulation pattern in the Mediterranean probably has existed since at least the Oligocene (Fig. 1.5). As result, an extensive quantity of deltaic sediments, debouched into the sea since the Oligocene, similarly was transported eastward by longshore current and deposited. Such a circulation pattern would have had an important bearing on the Neogene stratigraphy of the eastern Mediterranean and, in particular, on the Miocene in the Gulf of Suez.

Because the western flanks of the basins were filled as a result of Oligocene and early Miocene sedimentation, a slight shift in depocenters probably occurred during the early Miocene (Fig. 1.6), when, in the east, a major event took place; the Gulf of Suez was opened as an arm or trough of the Mediterranean Sea (Fig. 1.7). A combination of west-to-east longshore currents and orientation or position of shorelines may have diverted and funneled water and sediments into the Gulf of Suez. This could have resulted in the transport of early Miocene sediments from the Mediterranean to be deposited in the Gulf of Suez.

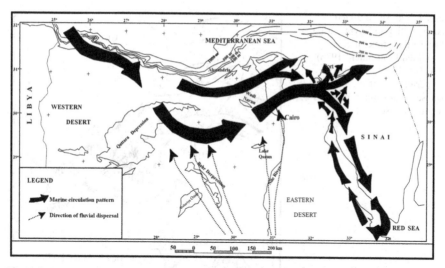

Fig. 1.5 Hypothetical circulation pattern of marine waters between Mediterranean and Gulf of Suez during Oligocene and Miocene time (modified from Salem 1976). AAPG©[1976], "reprinted by permission of the AAPG whose permission is required for further use"

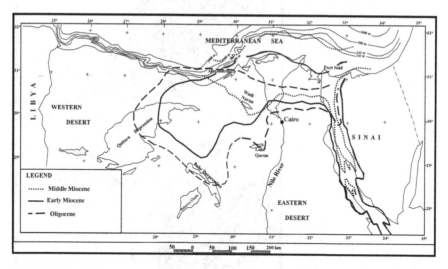

Fig. 1.6 Area with maximum terrigenous sedimentation (exclusive of carbonate deposits) during Oligocene and Miocene time (modified from Salem 1976). AAPG©[1976], "reprinted by permission of the AAPG whose permission is required for further use"

Middle Miocene: During this period the River Nile, as a result of subsidence and faulting, began to flow along a course very similar to the modern Nile. A thick deltaic sequence accumulated near the site of the present Nile delta. However, based on the proposed long shore-circulation patterns, some destructive type of

Fig. 1.7 Basin and land
configuration in Egypt during
late Tertiary time after
(modified from Salem 1976).
AAPG©[1976], "reprinted
by permission of the AAPG
whose permission is required
for further use"

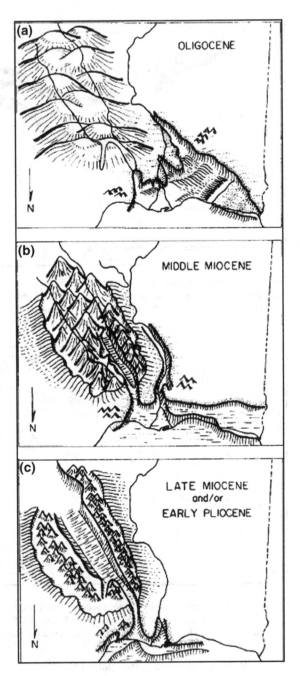

delta can be expected. The types of deltas built and the associated environments during middle Miocene time cannot be determined. West of the middle Miocene delta, the Marmarica Limestone banks continued to advance eastward over abandoned and foundering deltas (Salem 1976).

References

Abd-Alla A (2001) Sedimentology, mineralogy and depositional environments of the Moghra formation (Lower Miocene) at Minquar Abu Duweis, Qattara Depression, Egypt. Egypt J Geol 45:353–369

Abdallah A (1966) Stratigraphy and structure of a portion in the North Western Desert of Egypt (El Alamien-Dabaa-Qattara-Moghra areas) with reference to its economic potentialities. Geol Serv Egypt 45:1–19

Abdallah MA (1996) Sedimentary and stratigraphy of the exposed Neogene rocks in Wadi El-Natrun area, Western Desert, Egypt. Mansoura University, Mansoura, p 149

Abdeldayem AL (1996) Paleomagnetism of some Miocene rocks, Qattara depression, Western Desert, Egypt. J Afr Earth Sci 22:525–533

Abu-Zeid M, Al-Wakeel M (1992) Petrology, sedimentation and diagensis of the Moghra formation, Northern Qattara depression, Western Desert, Egypt. Egypt Mineral, Mineral Soc Egypt 4:213–237

Albritton CC, Brooks JE, Issawi B, Swedan A (1990) Origin of the Qattara depression, Egypt. Geol Soc Am Bull 102:952–960

Al-Wakeel MA (1989) Stratigraphical and sedimentological studies on the Miocene exposures around Moghra area Western Desert Egypt. [unpublished Master of Science thesis]: Faculty of science, Ain Shams University, Cairo, 247 p

Blanckenhorn M (1901) Neues zur Geologie und Palaeontologie Aegyptens: III: Das Miozän. Zeitschrift der Deutschen Geologischen Gesellschaft 53:52–132

Desio A (1928) Risultati scientifical della Miossione alla oasi di Giarabub.Fasc. 1: La morfologia. Soc Geol Italy, pp 1–82

El Bassyony AA (1971) Report on the geology of the Red Sea coastal strip between Lat. 25° 24′ 30″ and 23° 30′ 00″. Geol Surv U.A.R., inter. report, Doc. no. 35/71

El-Khashab B (1977) Some studies on Egyptian vertebrates fossils. Geol Surv Egypt 62:39

El-Khoriby EM (2004) Mechanism and factors affecting on the petrification of the wood forest, North Moghra Oasis, Western Desert, Egypt. (Abstract). 6 th intern. Conf. on Geochemistry, Alex. Univ. Egypt 6:255–270

Fisher WL, Brown LF Jr, Scott AJ, McGowen JH (1969) Delta systems in the exploration for oil and gas-research colloquium, syllabus. University of Texas at Austin, Bur Econ Geol, pp 1–78

Fourtau R (1918) Contribution à l'étude des Vertébrés miocènes de l'Egypte. Survey Department

Fourtau D (1920) Contribution a l'etude des vertébrés miocènes de l'Egypte. Govern-ment Press, Cairo, p 122

Hafez NAA (1992) Sedimentological and Mineralogical studies on the Miocene succession, Qaret El Kish, North Qattara Depression, Westrn Desert, Egypt. Al-Azhar Bull Sci 3:561–578

Hantar G (1990) North Western Desert. In: Said R (Ed) The geology of Egypt. AA Balkema, Rotterdam, pp 293–319

Haq B, Hardenbol J, Vail P (1987) Chronology of fluctuating sea levels since the Triassic. Science 235:1156

Haq B, Hardenbol J, Vail P (1988) Mesozoic and Cenozoic chronostratigraphy and cycles of sea-level change. In: Wilgus CK, Hastings BS, Kendall CGSC, Posamentier HW, Ross CA, Van Wagoner JC (Eds) Sea-level changes: an integrated approach: Society of Economic Paleontologists and Mineralogists Special Publication, SEPM Special Publication, pp 71–108

Howarth R, Mcarthur J (1997) Statistics for strontium isotope stratigraphy: a robust LOWESS fit to the marine Sr-isotope curve for 0 to 206 Ma, with look-up table for derivation of numeric age. J Geol 105:441–456

Joint-Venture Qattara (1981) Study Qattara depression-topography, regional geology, and hydrology. Qattara Project Authority, Ministry of Electricity, Egypt, p 129

Khaleifa MA, Abu Zeid K (1985) Miocene environmental models in the area northwest of W. El Natrun, Western Desert, Egypt (Abstract-Geology)

Marzouk I (1970) Rock stratigraphy and oil potentialities of the Oligocene and Miocene in the Western desert of Egypt. 7th Arab Petrol Congr 54:1–37

Miller E, Simons E (1996) Relationships between the mammalian fauna from Wadi Moghara, Qattara Depression, Egypt, and other early Miocene faunas. In: Proceedings of the Egyptian Geological Survey Centennial Conference, Nov 1996, pp 547–580

Misak RF (1979) Geology of the area between the Moghra Oasis and the Mediterranean Sea, Western Desert, Egypt [unpublished PH.D thesis]. Faculty of Science, Ain Shams University, Cairo

Norton P (1967) Rock stratigraphic nomenclature of the Western Desert, Egypt. General Petroleum Corporation of Egypt internal report, p 557

Omara S, Ouda K (1972) Review of the lithostratigraphy of the Oligocene and Miocene in the northern Western Desert. 8th Arab Perol. Cong. Algiers

Omara S, Sanad S (1975) Rock stratigraphy and structural feature of the area between Wadi El Natrun and the Moghra depression (Western Desert). Egypt Geol Jahrb 16:45–73

Oslick J, Miller K, Feigenson M, Wright J (1994) Oligocene-Miocene strontium isotopes: stratigraphic revisions and correlations to an inferred glacioeustatic record. Paleoceanography 9:427–443

Pekar S, Christie-Blick N (2008) Resolving apparent conflicts between oceanographic and Antarctic climate records and evidence for a decrease in pCO2 during the Oligocene through early Miocene (34–16 Ma). Palaeogeogr Palaeoclimatol Palaeoecol 260:41–49

Philip G, Darwishe M (1973) Mechanical analysis and mineral composition of the of the sandy sediments of the Moghra Formation in south of Alamein Area Western Desert, Egypt. Bull Fac Sci 6:507–531

Philip G, Abdallah AM, Darwishe M (1973) Petrographic studies on the Upper Tertiary rocks in the South of Alamein Area North Western Desert, Egypt. Bull Fac Sci 6:485–506

Pickford M (1991) Revision of the neogene Anthracotheriidae of Africa. In: Salem MJ, Busrewil MT (eds) The geology of Libya. Academic Press, New York, pp 1491–1525

Rizk Z, Davis A (1991) Impact of the proposed Qattara Reservoir on the Moghra aquifer of northwestern Egypt. Ground Water 29:232–238

Sadek A (1968) Contribution to the Miocene stratigraphy of Egypt by means of miogypsinids. In: Proceedings of the 3rd Afr Micropaleontol Colloq, Cairo, pp 509–514

Said R (1960) New light on the origin of the Qattara Depression. Bulletin de la Société de Géographie d'Égypte XXXIII:37–44

Said R (1961a) Tectonic framework of Egypt and its influence on distribution of foraminifera. AAPG Bull 45:198–218

Said R (1961b) Über das Miocän in der WestlichenWuste Ägyptens. Geol Jahrb 45:80

Said R (1962a) The geology of Egypt. Elsevier Publishing Company, Amsterdam 370 p

Said R (1962b) Das Miozän in der Westlichen Wuste Ägyptens. Geologisches Jahrbuch 80:349–366

Said R (1971) Explanatory notes to accompany the geological map of Egypt. Geol Surv Egypt 56:123

Said R (1990) The geology of Egypt. AA Balkema, Netherlands

Salem R (1976) Evolution of Eocene-Miocene sedimentation patterns in parts of northern Egypt. AAPG Bull 60:34–64

Sanders WJ, Miller ER (2002) New proboscideans from the early Miocene of Wadi Moghara, Egypt. J Vertebr Paleontol 22(2):388–404

Scruton PC (1960) Delta building and the deltaic sequence. In: Shepard FP, Phleger FB, Van Andel TH (Eds) Recent sediments, northwest Gulf of Mexico, Tulsa. American Association of Petroleum Geologists, pp 82–102

Sharaf EF (1995) Sedimentology and mineralogy of some Miocene deposits around Qattara Depression Western Desert Egypt. Mansoura University, Mansoura, p 125

Shata A (1955) An introductory note in the Geology of the Northern portion of the Western Desert of Egypt. Bull Inst Desert Egypt 5:96–106

Stratigraphic Sub-Committe of the National Committe Geological Sciences (1974) Miocene rock stratigraphy of Egypt. Egypt J Geol (Cairo, National Information and Documentation Centre (NIDOC)) 18:1–69

Chapter 2
Stratigraphy

Abstract The Lower Miocene Moghra Formation, northwestern Egypt was studied for its stratigraphy and geo-chronology. The Moghra Formation was divided into seventeen units according to lithology. The first unit is represented by marine fine-grained siliciclastic deposits with *Ophiomorpha* trace fossils (Unit I), and this is overlain by Unit II that consists of shallow marine coarse-grained siliciclastic deposits. This unit is rich in vertebrate fossil fragments. Unit III is composed of fine to coarse-grained siliciclastc deposits with *Ophiomorpha* trace fossils. Unit IV consists of a thick shale section with erosive base and variable thickness. The sediments of the overlying Unit V consist of three bioturbated sandstones beds. Unit VI is based by a major erosional surface and consists of fluvial-tidal sediments rich in vertebrates and silicified trunks. This unit is similar to Unit II in composition. Unit VI is overlain by unit VII, which is represented by sand-shale intercalation (sand dominated) and becomes more shaley upwards with *Ophiomorpha* and *Thalassinoid* trace fossils. Units VIII, Unit X, XII, XIV and XVI are similar to Unit VI in their composition. Unit IX is based by heterolithic strata with burrows and topped by marine shales towards the east. Units XI, XIII, XV are represented by calcareous beds and rich in *Ophiomorpha* and bioturbation ichnofacies. Unit XVII is represented by fossiliferous limestone and shale. Strontium isotope analysis of macrofossil fragments within Moghra Formation has provided a geochronology for the section and established a correlation with the global time-scale. Strontium isotope ratios of macrofossils are consistent and indicate an age for the Moghra Formation ranging from 20.5 Ma at the base to 17 Ma at the top, placing most of the study area within the Burdigalian.

2.1 Introduction

The Miocene stratigraphic work of previous authors in the Western Desert of Egypt has been discussed in Chap. 1. The present chapter deals with the detailed stratigraphy of the Lower Miocene Moghra Formation in the study area.

S. M. Hassan, *Sequence Stratigraphy of the Lower Miocene Moghra Formation*
in the Qattara Depression, North Western Desert, Egypt, SpringerBriefs in Earth Sciences,
DOI: 10.1007/978-3-319-00330-6_2, © The Author(s) 2013

The present study in the area of Moghra draws upon the principles of sequence stratigraphy to provide an integrative technique for forming and testing correlation based hypotheses.

Thirty seven GPS-based geological profiles, in the northeastern part of the Qattara depression, were measured for the lithological and sedimentological (both vertical and horizontal) characteristics of the exposed rocks as well as the vertebrate and invertebrate fossil content. From the thirty seven geological profiles around sixteen sections were made. Several locations between sections were described for the lateral correlation and variation in the lithology and tracing the depositional facies and bounding surfaces (Appendix). The sections from west to the east as follows: 22, 25, 24, 23, 21, 20, 9, 7, 8, 10, 1, 2, 3, 4, 6 & 5 (Fig. 2.1). The aim is to distinguish the main diagnostic characteristics through the succession with regard to sedimentary features and environments, diagenetic phenomena, thickness, boundaries and unit distribution. The objective also is to elucidate the environmental setting of the different sedimentary units within the measured profiles. This helps evaluate the sequences stratigraphic analysis, considered as one of the main goals in the present study. The description of the different lithologic units, thicknesses and sedimentary structures are presented here under. Symbols for lithology and sedimentary structures in these figures are supplied in the legend figure in appendix.

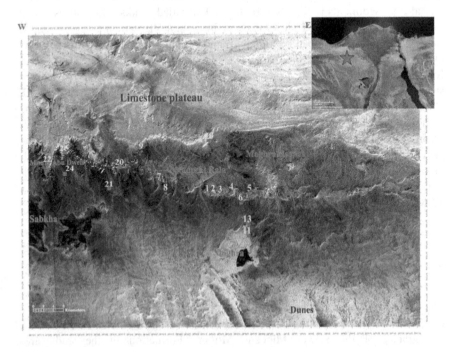

Fig. 2.1 Staellite image of the study area, showing the measured section from East to the West

2.2 Stratigraphy of the Area

The studied successions could be subdivided according to the dominant lithology into repeated cycles of intercalated sandstone and mudstone. Most cycles are erosionally-based by cross-bedded sandstone and terminated by *Ophiomorpha* sand and other marine strata. The studied succession was subdivided into sixteen units according to lithology. The first one is represented by marine fine siliciclastics deposits with *Ophiomorpha* trace fossils (unit I), which is overlain by unit II that consists of shallow marine coarse siliciclastic deposits (Unit II). This unit is rich in vertebrate fossil fragments. Unit III is composed of fine to coarse siliclclastc deposits with *Ophiomorpha* trace fossils. Unit IV is represented by thick shale section with erosive base and variable thickness. The sediments Unit V, is recorded overlying unit IV and consists of three bioturbated sandstones. Unit VI is based by erosional surface and consists of fluvial input sediments rich in vertebrates and silicified trunks. This unit is similar to unit II in composition. Unit VI is overlain by unit VII, which is represented by sand shale intercalation with the sand dominated and become more shaley upward with *Ophiomorpha* and *Thalassinoid* trace fossils. Units VIII, Unit X, XII, XIV and XVI are similar to unit VI in their composition. Unit IX is based by heterolithic bed with burrows and topped by marine shale bed towards east. Units XI, XIII, XV are represented by calcareous beds and rich in *Ophiomorpha* and bioturbation. However, these units have not been discussed in details because they have now been superseded by sequence stratigraphy subdivision (more details of sequence subdivision in Chap. 5). These lithostratigraphic units however, have been useful for the preliminary field correlation before constructing the sequence stratigraphic model.

The following table gives an extended summary for the main stratigraphic units of the northeastern part of Moghra Depression (Table 2.1).

2.3 Geochronology

There is a lack of suitable material for absolute dating of the critical mid Cenozoic Moghra Formation by conventional methods such as K/Ar or ^{40}Ar/^{39}Ar. However, indirect dating using strontium isotope stratigraphy is possible. Precise and detailed data on the variability of strontium isotopes with time in the world's oceans is now known for much of Phanerozoic time (see Howarth and Mcarthur 1997) and local curves for specific stratigraphic sections are increasingly used to infer absolute ages (e.g., Hurst 1986; Mckenzie et al. 1988; Rundberg and Smalley 1989; Smalley et al. 1986; Whitford et al. 1996). For the mid Cenozoic in particular, the strontium isotope sea-water curve changes rapidly with time, and so is particularly suitable for geochronology. In this time interval, resolution of stage boundaries is better than 0.5 m.y. (Howarth and Mcarthur 1997; Oslick et al. 1994), making the method a powerful tool for improving correlation between biostratigraphic and chronostratigraphic timescales (Graham et al. 2000).

Table 2.1 The stratigraphic classification of the studied sections

Section	Thickness (m)	WP	Location	Units/Thick (m)	Field description
1	77	003		V, 19.5	Bioturbated bed + shell (pelecypods).
				IV, 8	Shale + plant remains
				III, 10	Burrowed sandstone + *Ophiomorpha*
				II, 29.5	Sand, vertebrates, wood.
				I, 10.1	*Ophiomorpha* sandstone
Aux	38.5		30° 21.255'N 28° 52.195'E	V, 10.5	Highly bioturbated sand with thick *Thalassinoides* rather than *Ophiomorpha* + sandstone with ripples & flasers + fossiliferous sandstone with pelecypods, shale layers
				IV, 22	Black shale with slumping structure.
				II, 5	Fluvial sand, vertbrates
2	41.8	6	30° 21.263'N 28° 52.321'E	II, 32	Fluvial sand, vertbrates
				I, 12.2	Burrowed sand shale intercalated.
3	14.7	020 Top 001 Base	30° 20.911'N 28° 53.625'E 30° 20.909'N 28° 53.625'E	II, 7.7	Sandstone
				I, 7	Sandstone, burrowed, gypsum
4	111.3	013	30° 21.039'N 28° 55.130'E	X, 20	Sandstone, vertebrates and trunks
				IX, 3.5	Sand- shale intercalations
				VIII, 8.5	Sandstone, vertebrates and trunks
				VII, 6	Wavy and ripple laminated heterolithic bed + siltstone with ferruginous hard crust.
				VI, 10.3	Coarse sand grain with silicified wood fragments and bone fragments.
				V, 16.5	Fine calcareous sand, highly bioturbated with *Thalassionoid*.
				IV, 12.15	Fine sand + lag with vertebrates + cross bedded sand + ripple and flaser sand intercalated with shale.
				III, 5.6	*Ophiomorpha* and *Thalassinoid* medium sand + laminted shale + heterolithic bed.
				II, 15	Sandstone, vertebrates and trunks
				I, 13.6	Sandstone, shale with *Ophiomorpha*
5	92		30° 21.173'N 28° 57.893'E	XII, 15	Fining up ward cross-bedded sandstone + roots and topped by Marmarica limestone.
				XI, 10	Calcareous cross-bedded sandstone with large root cast.
				X, 25	Slope forming, cross-bedded large scale fining upward sandstone.
				IX, 13.7	Calcareous sandstone with *Ophiomorpha* + micaceous sandstone
				VIII, 19	Sandstone with vertebrates + micaceous sandstone
				VII, 3.3	Shale + flaser and laminted sandstone.
				VI, 8.5	Fining upward sandstone, cross bedded + trunks

(continued)

Table 2.1 (continued)

Section	Thickness (m)	WP	Location	Units/Thick (m)		Field description
5'	15	63?	30° 21.541'N 28° 58.448'E	XVII	15	Cross-bedded fine sands with mangroves + thin laminated shale + hard dolomitic bioturbated.
6	163.5	42	30° 20.683'N 28° 56.697'E	X	18	Trough filling cross-bedded sandstone + silicified trunks in the base and bioturbated and botridal sand in the top + dolomite
				IX	2	Thin laminated, ripple silt with fine sand
				VIII	53	Slope forming, fine to medium cross-beded sans with 5 erosional surfaces marked by silicfied wood and vertebrates (coprolites) + mud shale intercalation in the top.
				VII	14	Highly bioturbated (honycamb) and homogenized sand + micaceous sand.
				VI	33	Slope forming cross-bedded coarse to medium sand with large trunks.
				V	12.1	Slope forming cross-bedded medium to fine grained sandstone with muddrapes and burrowed + hard mudstone + fractured mudstone and ferruginated
				IV	10	Cross-bedded sandstone mostly trough, with small scale tabular cross-bedding and silicified palm tree.
				III	16.8	Shale with gypsum and plantremains + *Ophiomorha* fine sand_channel fill cross-bedding sandstone + *Ophiomorha* sand.
				II	11	Tabular cross-bedded sand + trough cross bedding medium to fine sand + lag with vertebrates + trough cross-bedded sand.
				I	2	Highly bioturbated fine sand.
7	92.5	76	30° 22.389'N 28° 46.067'E	VI	12	Slope forming, medium to coarse cross-bedded sand with several erosional surfaces marked by silicfied wood and vertebrates.
				V	8	Slope forming, medium to coarse cross-bedded sand silicified wood and vertebrates.
				IV	4	Sand-shale intercalations, the sand rich in glauconite pellets and fissile shale.
				III	19.7	Fill structure, thin laminted heterolithic bed + burrowed sand + sand shale interclation + heterolithic bed.
				II	17.6	Slope forming, medium to coarse cross-bedded sand with three erosional surfaces marked by silicfied wood and vertebrates.
				I	6.6	Fine sand with mud drapes.
					15.3	Sand-shale intercalations, highly burrowed sand + shale with plant remains.
					8	Highly bioturbated heterolithic beds + micaceous sand with mud drapes.
8	28.5		30° 21.123'N 28° 47.562'E	II	10	Sandstone, vertebrates, wood
				I	18.2	Sand, shale intercalated, *Ophiomorpha*
9	86.5	69	30° 25.312'N 28° 45.004'E	XVII	22	Fossiliferous limestone, shale
				XVI	20	Sandstone, vertebrates, wood
				XV	21.5	Fossiliferous limestone
				XIII	24	Sand-shale intercalations
10	36	66	30° 26.294'N 28° 52.41'E	XVII	36	Sandstone, fossiliferous limestone

(continued)

Table 2.1 (continued)

Section	Thickness (m)	WP	Location	Units/Thick (m)		Field description
11&12	11.1	94	30° 17.527'N 28° 57.040'E	Zero?	11.1	Cross-bedded sand with small burrows + fractured claystone kaolinitic with small burrows + sandstone with vertical burrows + cross-bedded sand with erosive base + sand shale intercalated + cross-bedd sand + mudstone with burrowed + sand shale interclation + shale with burrows
13	26.3	95	30° 18'101'N 28° 57.539'E	Zero?	26.3	Sand with silicified + siltstone + cross-bedded sand + shale non fissile + sand with clay intercalation + sandstone with bone fragment + crust + cross-bedded with vertebrates and bone fragments.
20	136.2	181	30° 24.224'N	XVI	19	Sandstone, wood, vertebrates
		27	28° 42.844'E	XV	29	Calcareous Sandstone, shell fragments
				XIV	9	Sandstone
				XIII	4.5	Sand-shale intercalation
				XII	8.5	Sandstone
				XI	1.5	Calcareous sandstone, shell fragments, *Ophiomorpha*
				X	8.5	Sandstone, wood, vertebrates
				IX	22	Calcareous Sandstone, *Ophiomorpha*, shell fragments
				VIII	11.5	Sandstone, wood, vertebrates
				VII	5.5	Sand-shale intercalation
21	230.15		30° 22.87'N 28° 41.135'E	XVI	50	Sandstone, wood, vertebrates
				XV	12.5	Sand-shale intercalation
				XIV	9	Sandstone
				XIII	15.5	Sand-shale intercalation, burrowed
				XII	10	Sandstone, wood, vertebrates
				XI	35	Homogenized cross-bedded sand.
				X	25	Sandstone, wood, vertebrates
				IX	9	Sand-shale intercalation, sand lenses
				VIII	34.5	Sandstone, vertebrates
				VII	3.5	Sandstone, *Ophiomorpha, Thalassinoid*
				VI	8	Sandstone, wood, vertebrates
				V	2.5	Sandstone, *Ophiomorpha*
				IV	3	Sand-shale intercalation
				III	3	Sandstone, *Ophiomorpha*
				II	7.5	Cross-bedded sandstone
				I	15.5	Sand-shale intercalation

(continued)

Table 2.1 (continued)

Section	Thickness (m)	WP	Location	Units/Thick (m)		Field description
22	171.5		30° 24.662'N	XV	3	Calcareous sandstone
			28° 33.836'E	XIV	26.5	Sandstone, wood, vertebrates
				XIII	30.5	Calcareous sandstone, sand-shale intercalated, *Ophiomorpha*
				XII	19.5	Sandstone, bones, *Ophiomorpha*
				XI	26.5	Calcareous sandstone, *Ophiomorpha, Limestone*
				X	20	Sandstone, wood, vertebrates
				IX	2	Sand-shale intercalation, burrowed
				VIII	8	Sandstone, gypsum at the base
				VII	20.5	Sand-shale intercalation, burrowed (*Ophiomorpha*)
				VI	7.5	Sandstone
23	13	197	30° 21.66316	?	13	Calcareous sandstone
			28° 37.10199			
24	61	204	30° 24.89'N	XI	7	Calcareous sandstone
			28° 35.114'E	IX	9.5	Sand-shale intercalation, *Ophiomorpha, Sandstone*
				VIII	22	Sandstone
				VII	22.5	Sand-shale intercalation, *Ophiomorpha*
25	88.5	205	30° 26.14962	XIII	41.3	Fossiliferous limestone_shale-sand intercalation + cross-bedded sandstone
			28° 34.92005	XII	30	Slope forming cross-bedded sandstone
				XI	18	Shale-sand intercalation + Fossiliferous limestone
26	30	213&214		Zero	30	Fine calcareous sandstone with largetrunks + shale + fine sandstone
30	87		30° 20.77512	VI	12.5	Sandstone, wood, vertebrates
			28° 52.90753	V	27	Sandstone, *Ophiomorpha*
				IV	4	Mudstone with plant remains
				III	4	Cross-bedded sandstone
				II	6	Sandstone, wood, vertebrates
				I	31.5	Shale, sandstone, wood, vertebrates
32	33		30° 26	XVII	33	Sand-shale intercalation, fossiliferous limestone
			51.421			
			28° 46			
			51.961			

*Zero is a unit candidate to be below unit I

2.3.1 Previous Work and Discussion

The major difficulty in assessing the age of Moghra is that the site is not associated with volcanic deposits, making it impossible to date the Moghra mammals radiometrically. Some of the previous works considered the only way to assess the age of the Moghra mammals is by faunal correlation with fossil sites that have absolute dates. In East Africa, a time successive sequence of radiometric dates has been developed for a number of localities ranging across the early and middle Miocene, and the age of the Moghra fossil mammals can be estimated by comparison of the Moghra fauna with that of East Africa as indicated by Miller and Simons (1996). They argued that the most conservative estimate for the age of Moghara is 18–17 Ma, approximately the same age as the Rusinga (Hiwegi) fauna. In addition, relatively rarer faunal elements shared between Moghara and Napak, but not with Zelten in Libya, suggest that Moghara is older than Zelten. The same authors stated, however, that the evidence that Moghara may be as old as Napak (ca. 19 Ma) is not compelling as almost all genera shared between Moghara and Napak are also found at Rusinga (ca. 18–17 Ma). These findings are in general agreement with those reached by Pickford (1991) concerned with the age of Zelten, and confirm the hypothesis of Geraads (1987) that Moghra is older than Zelten. They indicated that the Moghara mammals are probably about 18–17 Ma, and have their closest biogeographic affinities with certain East African sites. In fact, it appears that the Moghara fauna is more similar to the mammals from a number of East African sites than it is to the fauna from Gabal Zelten, Libya.

McCrossin (2008) concluded that the assessment of the Jabal Zaltan and Moghra faunas indicates that previous attempts at biochronologic correlation oversimplified the span of time represented by these deposits. Rather than being roughly equivalent to Maboko (ca. 15–16 Ma), the mammal faunas of Jabal Zaltan extend for long periods of time, from the terminal Oligocene or basal Miocene (ca. 22–26 Ma?) in the northern reaches of the Marada Formation to the middle-later part of the Middle Miocene (ca. 12–15 Ma) near Wadi Shatirat. He also mentioned that mammal fossils from Moghara date not only from the later part of the early Miocene (ca. 17 Ma) but also from the early part of the middle Miocene (ca. 15 Ma). Contrary to widely held opinion, the cercopithecoid from Gabal Zelten is more primitive (and probably, therefore, more ancient) than *Prohylobates tandyi* from Moghara. Reassessment of the mammal faunas of Gabal Zelten and Moghara demonstrates a substantial degree of North African zoogeographic provincialism, together with connections to sub-Saharan Africa and Eurasia.

Others works considered the geochronology of the Neogene is based, to a large extent, on paleontological as well as stratigraphical evidence. The large collection of macroinvertebrates recorded from the Miocene of Egypt (Blanckenhorn 1900; Blanckenhorn 1901; Fourtau 1920; Sadek 1968; Said and Yallouze 1955, etc.) has not been successfully used to zone the Miocene rocks. Mention has frequently been made of the cephalopod *Aturia aturi* as an index of the Langhian (Said 1990). However, the present study succeeded in using macroinvertbebrates to date the

Miocene of Moghra Formation within Burdigalian. Agree with Abdallah (1966) mentioned that the Moghra Formation is of Burdigalian age and the lowermost Lower Miocene (Aquitanian) is not found. Furthermore, Upper Miocene (post— Tortonian-Helvetian) is absent also.

For earlier studies involving the ages of Moghra and Zelten based on previous radiometric information or correlation of marine (foraminifera) or land animals see Arambourg (1963); Bernor (1984); Desio (1935); Hamilton (1973a), (b); Harris (1973); Hoojier (1978); Pickford (1981), (1983), (1991); Savage (1967), 1969, 1971, 1990); Savage and Hamilton (1973); Savage and White (1965); Savage in Selley (1966); Tchernov et al. (1987); Thomas (1979), (1984); Wilkinson (1976); Van Couvering (1972); Van Couvering and Berggren (1977); Van Couvering and Van Couvering (1975).

2.3.2 Absolute Ages from Strontium Isotopes

2.3.2.1 General Principles

Dating marine sediments using strontium isotopes is based on the following assumptions and observations: (1) at any point in time, the $^{87}Sr/^{86}Sr$ ratio of sea-water is homogeneous throughout the world's oceans (Elderfield 1986; Faure 1986) because the oceanic residence time of strontium (24 m.y.) is much longer than the mixing time of the oceans (c. 0.001 m.y.) (Broecker and Peng 1982); (2) over geologic time, sea-water $^{87}Sr/^{86}Sr$ varies because of changes in the relative fluxes of strontium to the oceans from different sources (e.g., continental runoff, hydro-thermal outflow at mid-ocean ridges, diagenetic reflux from buried pore waters), each with its own characteristic $^{87}Sr/^{86}Sr$ ratio (Palmer and Edmond 1989; Richter et al. 1992). At steady state, these inputs are counteracted by removal of strontium via sedimentation, and exchange of radiogenic strontium in hydrothermal waters for that in basalts (Veizer 1989); and (3) strontium is removed from sea-water by co-precipitation in biogenic carbonate. The isotope composition of strontium in calcitic macrofossils is identical to that of the oceans in which they lived, provided that the macrofossils are well preserved and the effects of diagenesis are minimal (Richter and Depaolo 1987). Although planktic foraminifera are often used for calibration of sea-water curves, there is no convincing evidence that contempora-neous bottom-dwelling benthic foraminifera or macrofossils have significantly different $^{87}Sr/^{86}Sr$ values from their planktic cohabitors (Graham et al. 2000).

2.3.2.2 Methods

The sections of Moghra Formation have been well studied stratigraphically and have macrofossil faunas, providing excellent material for strontium isotope anal-ysis and for biostratigraphic control.

2.3.2.3 Fossil Collections

About 10 samples of fresh macrofossil shell material, mainly pectinids, Echino-dermata, and oysters, were collected from several previously well documented earliest Miocene sections within the Moghra Formation. In all cases the associated lithostratigraphic units (in this formation) bearing the fossils had previously been assigned to Lower Miocene. The fossils represent mainly isolated specimens from separate localities and different sections. Stratigraphic information for the analysed fossil collections is summarized in Table 2.2.

Table 2.2 Strontium isotope data and age calculations for macrofossils samples

Type of fossil	Sample No.	Unit	$Sr^{87/86}$	Precision	Age (Ma)
Mollusca (Pelecypoda)	(10-7)	XVII	0.708658	0.000018	17
Pecten (N)	(10-6-2)	XVII	0.708700	0.000010	16.5
Echinodermata	(10-3)	XVII	0.708637	0.000010	17.1
Echinodermata	(9-1-2)	XIII	0.708615	0.000010	17.4
Mollusca (Pelecypoda)	(20-9-1)	XV	0.708525	0.000018	18.2
Echinodermata	(22-12)	XI	0.708656	0.000011	18
Mollusca (Oyster)	(Ox2)	V	0.708534	0.000010	18.2
Mollusca (Oyster)	(1-2-5-1)	II	0.708170	0.000010	23
Mollusca (Oyster)	(3-2-2)	II	0.708445	0.000018	19.6
Mollusca (Pelecypoda)	(21-1-4)	I	0.708410	0.000018	21

Analyses undertaken at Lamont Doherty Earth Observatory Lab, Columbia University, New York.

2.3.2.4 Analytical Methods

Macrofossil material (mainly pectinid mollusks, Oyster and Echinodermata) were prepared. From the original suite of fossil samples, 10 were selected for Sr isotope analysis, from different localities. After careful cleaning to remove surficial impurities or rock matrix, the shells were powdered and small (30–50 mg) aliquots were leached in cold 1 M acetic acid (Bailey et al. 2000). Acetic acid was used for sample dissolution in order to minimize the extraction of strontium from dolomite, clays, and other terrigenous material (Depaolo 1986).

Sr was extracted from the leachates using a single pass over small (0.1 ml) beds of EICHROM™ Sr resin. Strontium isotope analysis of macrofossil material was carried out on a VG sector 54 multi-collector thermal ionization mass spectrom-etry (TIMS) at Lamont Doherty Earth Observatory Lab (Fig. 2.2), Columbia University, New York. General methods are described in Bailey et al. (2000).

Fig. 2.2 Mass spectrometer at Lamont Doherty earth observatory lab, Columbia university, New York

2.3.2.5 Results and Discussion

Macrofossil samples show a coherent pattern of slightly increasing strontium isotope ratios with stratigraphic height, indicating an upwards younging direction (Fig. 2.3) as seen in Hodell et al. 1990 (Fig. 2.4).

Macrofossil samples have significantly high strontium isotope ratios ranging from 0.708410 ± 0.000018 to 0.708658 ± 0.000018 (Table 2.2). There is a smooth and consistent increase in $^{87}Sr/^{86}Sr$ up through the section, the basal two samples having a mean of 0.7084275 (20.05 $\pm$? Ma), the middle three samples

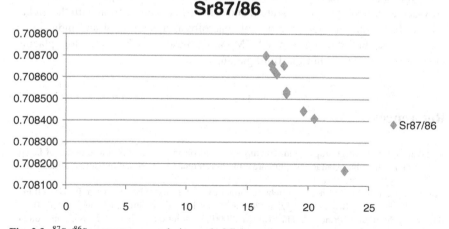

Fig. 2.3 $^{87}Sr/^{86}Sr$ seawater curve during early Miocene

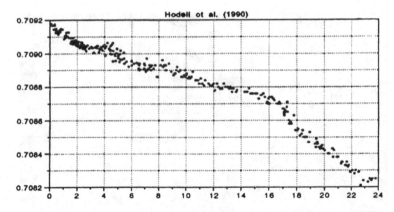

Fig. 2.4 Variation in the stronium isotopic composition of seawater on a progressively expanding timescale during Neogene (modified from Hodell et al. 1990)

having a mean of 0.7085716 (18.3 ± ? Ma) and the top three samples a mean of 0.708636 (17.16 ± 0.? Ma) Table 2.2.

The ^{87}Sr/^{86}Sr of two samples 1-2-5-1 and 10-6-2 are significantly different from adjacent samples, and have been excluded as an outlier. The sample 1-2-5-1 is oyster and sample 10-6-2 is pectin, thick-shelled pectin distinctly different from the thin-walled pectinids which comprise the other samples. Two excluded samples have lower ^{87}Sr/^{86}Sr suggesting they could be older, reworked from lower down in the succession, or they were originally aragonitic and their original ^{87}Sr/^{86}Sr have been altered by neomorphism and/or diagenesis.

2.3.2.6 Summary

Strontium isotope analysis of macrofossil fragments from Moghra Formation has provided a chronology for the section and established a correlation with the global time-scale. Strontium isotope ratios of macrofossil are identical and indicate an age for the section ranging from 20.5 Ma at the base to 17 Ma at the top, placing most of the section within the Burdigalian.

References

Abdallah A (1966) Stratigraphy and structure of a portion in the North Western desert of Egypt (El Alamein-Dabaa-Qattara-Moghra areas) with reference to its economic potentialities. Geol Serv Egypt, Paper, 45:1–19

Arambourg C (1963) Continental vertebrate faunas of the tertiary of North Africa. In: Howell FC, Bourliere F (ed) African ecology and human evolution. Aldine Press, Chicago, p 55–60

Bailey T, Mcarthur J, Prince H, Thirlwall M (2000) Dissolution methods for strontium isotope stratigraphy: whole rock analysis. Chem Geol 167:313–319

Bernor R (1984) A zoogeographic theater and biochronologic play: the time-biofacies phenomena of Eurasian and African Miocene mammal provinces. Paléobiologie continentale 14:121–142

Blanckenhorn M (1900) Neues zur Geologic and Palaontologie Aegyptens 11, Das Palaeogen (Eocan und Oligocan). Deut Geol (Jesell. Zeitschr) 52:403–479

Blanckenhorn M (1901) Neues zur Geologie und Palaeontologie Aegyptens: III: Das Miozän. Zeitschrift der Deutschen Geologischen Gesellschaft 53:52–132

Broecker W, Peng T (1982) Tracers in the sea, vol. 690. Lamont-Doherty Geological Observatory, Palisades, New York

Depaolo D (1986) Detailed record of the neogene Sr isotopic evolution of seawater from DSDP site 590B. Geology 14:103–106

Desio A (1935) Missione scientifica della Reale Accademia d'Italia a Cufra (1931-IX); vol. I, Studi geologici sulla Cirenaica, sul deserto Libico, sulla Tripolitania e sul Fezzan orientali

Elderfield H (1986) Strontium isotope stratigraphy. Palaeogeogr Palaeoclimatol Palaeoecol 57:71–90

Faure G (1986) Principles of isotope geology. Wiley, New York, p 589

Fourtau D (1920) Contribution a l'etude des vertébrés miocènes de l'Egypte. Government Press, Cairo, p 122

Geraads D (1987) Dating the northern African cercopithecid fossil record. Human Evolution 2:19–27

Graham I, Morgans H, Waghorn D, Trotter J, Whitford D (2000) Strontium isotope stratigraphy of the Oligocene-Miocene Otekaike Limestone (Trig Z section) in Southern New Zealand: age of the Duntroonian/Waitakian stage boundary. N Z J Geol Geophys 43:335–348

Hamilton W (1973a) The lower Miocene ruminants of Gebel Zelten, Libya. Br Mus (Nat Hist) 21:73–150

Hamilton W (1973b) North African lower Miocene rhinoceroses. Br Mus (Nat Hist) 24:351–395

Harris J (1973) Prodeinotherium from Gebel Zelten, Libya. Br Mus (Nat Hist) 23:285–348

Hodell DA, Mead GA, Mueller PA (1990) Variations in the strontium isotopic composition of seawater (8 Ma to present): Implications for chemical weathering rates and dissolved fluxes to the oceans. Chem Geol 80:291–307

Hooijer DA (1978) Rhinocerotidae. In: Maglio VJ, Cooke HBS (eds) Evolution of African mammals. Harvard University Press, Cambridge, p 371–378

Howarth R, Mcarthur J (1997) Statistics for strontium isotope stratigraphy: a robust LOWESS fit to the marine Sr-isotope curve for 0 to 206 Ma, with look-up table for derivation of numeric age. J Geol 105:441–456

Hurst R (1986) Strontium isotopic chronostratigraphy and correlation of the Miocene monterey formation in the Ventura and Santa Maria basins of California. Geology 14:459–462

Mccrossin ML (2008) Biochronologic and Zoogeographic relationships of early-middle Miocene mammals from Jabal Zaltan (Libya) and Moghara (Egypt). Geol East Libya 3:267–290

Mckenzie J, Hodell D, Mueller P, Mueller D (1988) Application of strontium isotopes to late Miocene-early Pliocene stratigraphy. Geology 16:1022–1025

Miller E, Simons E (1996) Relationships between the mammalian fauna from Wadi Moghara, Qattara depression, Egypt, and other early Miocene faunas. In: Proceedings of the Geological Survey of Egypt Centennial Conference, p 547–580

Oslick J, Miller K, Feigenson M, Wright J (1994) Oligocene–Miocene strontium isotopes: stratigraphic revisions and correlations to an inferred glacioeustatic record. Paleoceanography 9:427–443

Palmer M, Edmond J (1989) The strontium isotope budget of the modern ocean. Earth Planet Sci Lett 92:11–26

Pickford M (1981) Preliminary Miocene mammalian biostratigraphy for western Kenya. J Hum Evol 10:73–97

Pickford M (1983) Sequence and environments of the lower and middle Miocene hominoids of western Kenya. In: Ciochon R, Corruccini RS (eds) New interpretations of ape and human ancestry. Plenum Press, New York, p 421–439

Pickford M (1991) Biostratigraphic correlation of the middle Miocene mammal locality of Jabal
    Zaltan, Libya. In: Salem MJ, Busrewil MT (eds) The geology of Libya. Academic Press, New
    York, p 1483–1490
Richter F, Rowley D, Depaolo D (1992) Sr isotope evolution of seawater: the role of tectonics.
    Earth Planet Sci Lett 109:11–23
Richter FM, Depaolo DJ (1987) Diagenesis and Sr isotope evolution of sea-water using data from
    DSDP site 590B and 575. Earth Planet Sci Lett 90:382–394
Rundberg Y, Smalley P (1989) High-resolution dating of cenozoic sediments from northern
    North sea using 87Sr/86Sr stratigraphy. AAPG Bull 73:298–308
Sadek A (1968) Contribution to the Miocene stratigraphy of Egypt by means of miogypsinids.
    Proc 3rd Afr Micropaleontol Colloq Cairo 509–514
Said R (1990) The geology of Egypt. AA Balkema, Rotterdam
Said R, Yallouze M (1955) Miocene fauna from Gebel Oweibid, Egypt. Bull Fac Sci Cairo Univ
    33:61–81
Savage R (1967) Early Miocene mammal faunas of the Tethyan region. Syst Assoc Publ Lond
    7:247–282
Savage R (1969) Early tertiary mammal locality in southern Libya. Proc Geol Soc Lond
    1657:167–171
Savage R, Hamilton W (1973) Introduction to the Miocene mammal faunas of Gebel Zelten,
    Libya. Br Mus (Nat Hist) 22:515–527
Savage R, White M (1965) Two mammal faunas from the early tertiary of central Libya. Proc
    Geol Soc Lond 1623:89–91
Savage RJG (1971) Review of the fossil mammals of Libya. In: Gray C (ed) Symposium on the
    Geology of Libya. Faculty of Science, University of Libya, Tripoli, Libya, p 215–225
Savage RJG (1990) The African dimension in early Miocene mammal fauna. In: Lindsay EH (ed)
    European neogene mammal chronolgy. Plenum Press, New York, pp 587–599
Selley R (1966) The Miocene rocks of the Marada and Jebel Zeltan. A study of shoreline
    sedimentation. Geol Soc Lond 1623:89–91
Smalley P, Nordaa A, Raheim A (1986) Geochronology and paleothermometry of neogene
    sediments from the Vøring plateau using Sr C and O isotopes. Earth Planet Sci Lett
    78:368–378
Tchernov E, Ginsburg L, Tassy P, Goldsmith N (1987) Miocene mammals of the Negev (Israel).
    J Vertebr Paleontol 7:284–310
Thomas H (1979) Les bovide'nes Miocenes rifts est-africans: implications pale'obiographiques.
    Bull Soc Ge'ol Fr 21:295–299
Thomas H (1984) Les Bovidae (Artiodactyla:Mammalia) Mioce/ne du sous-continent indien de la
    pe/ninsule arabique et de l'Afrique:biostratigraphie, bioge/ographie et e/cologie. Palaeogeogr
    Palaeoclimatol Palaeoecol 45:251–299
Van Couvering J (1972) Radiometric calibration of the European neogene. In: Bishop W, Miller J
    (eds) Calibrarion of hominoid evolution. Scottish Academic Press, Edinburgh, p 247–272
Van Couvering J, Berggren W (1977) Biostratigraphical basis of the Neogene time scale. In:
    Kauffman EG, Hazel JE (eds) Concepts and methods in biostratigraphy. Drowden,
    Hutchinson & Ross, Pennsylvania, p 283–306
Van Couvering J, Van Couvering J (1975) African isolation and the Tethys seaway. In: VI th b,
    Congress regional committee on mediterrraen neogene stratigraphy, p 363–367
Veizer J (1989) Strontium isotopes in seawater through time. Annu Rev Earth Planet Sci
    17:141–167
Whitford D, Allan T, Andrew A, Craven S, Hamilton P, Korsch M, Trotter J, Valenti G (1996)
    Strontium isotope chronostratigraphy and geochemistry of the Darai limestone: Juha 1X Well,
    Papua New Guinea: Petroleum exploration and development in Papua New Guinea. In:
    Proceedings of the 3rd PNG petroleum convention, p 369–380
Wilkinson A (1976) The lower Miocene Suidae of Africa. Fossil Vertebr Afr 4:173–282

Chapter 3
Sedimentary Facies

Abstract An outcrop investigation of Moghra Formation (Lower Miocene, Burdigialian age) was carried out in northwestern Egypt. Eighteen detailed sedimentary measured sections, located slightly obliquely to the depositional strike were described. Emphasis was placed on lithofacies variations, interpretations of depositional settings and a depositional model was constructed. In the study area, The Moghra Formation, some 260 m thickn, consists of eight lithofacies associations: (1) Tide-influenced fluvial channel deposits; (2) Flat laminated sandflat deposits; (3) Outer estuary sand bar deposits (Tidal Channel and Tidal Bars); (4) Tidal flat deposits; (5) Bioturbated fossiliferous shelf sandstones; (6) Bioturbated fossiliferous shelf carbonates; (7) Coarsening upward deltaic deposits; and (8) Fining upward channel deposits. These eight lithofacies associations are grouped into three main depositional environments: (1) Transgressive tide-dominated estuaries, (2) Open Shelf and (3) Regressive tide-dominated deltas.

3.1 Introduction

This chapter aims to analyse and discriminate the different sedimentary facies constituting the studied Lower Miocene succession of Moghra Formation in the northern part of the western desert. The discrimination of the sedimentary facies associations is made on the basis of lithology, microfacies associations, sedimentary structures, depositional textures and faunal content. These criteria are integrated to create facies associations and appropriate depositional environments.

3.2 Facies Associations

Several localities in Moghra formation were studied in detail. The selection was based on outcrop quality, but location in relation to anticipated changes in stratigraphic architecture was also considered. Several vertical sections were

S. M. Hassan, *Sequence Stratigraphy of the Lower Miocene Moghra Formation in the Qattara Depression, North Western Desert, Egypt*, SpringerBriefs in Earth Sciences, DOI: 10.1007/978-3-319-00330-6_3, © The Author(s) 2013

measured within each locality, recording parameters such as grain size, sedimentary structures, color and bounding surfaces. The sections were sufficiently closely spaced that they could be correlated laterally by walking out surfaces and units. Each locality was photographed in detail to provide additional information for sedimentological and correlation purposes. Finally, the studied localities were compared and correlated. The study of 18 sedimentary sections located along the depositional strike reveals complex arrangements of facies associations and stratal architecture for the seventeen units described in the previous chapter. The deposits of the Moghra formations have been classified into twenty facies and grouped into eight facies associations. See Tables 3.1 and 3.2, for detailed descriptions and interpretations.

A complete transect through the Moghra succession shows the presence of three main depositional environments: (1) Transgressive tide dominated estuary, (2) Open Shelf and (3) Regressive tide-dominated delta succession (Table 3.1). These interpretations will be justified by argument below.

3.2.1 Tide-Dominated Estuary

The estuary environment of Moghra area contains two facies associations. These associations are interpreted as (a) axial, high-energy tidal channels with infilling bars/dunes that proximally become tidal-fluvial channels and (b) marginal, low-energy intertidal to supratidal flats (as is common in most modern estuaries), from near the mouth of the estuarine funnel to the landward limit tidal influence. These facies associations were reconstructed using sedimentary characteristics and dominant physical processes. These two estuarine depositional environments have then been divided into a series of subenvironments, following Dalrymple and Choi (2007). We note also that the boundaries between all of the following sub-environments are gradational and relatively arbitrary.

3.2.1.1 Axial Estuary Zone is Subdivided Into

a. Inner-most estuary, tide-influenced fluvial channels
b. Mid-estuary, upper-flow regime sand-flats
c. Outer-estuary tidal bars and channels
d. Tide-influenced fluvial channel deposits (FA1).

Channel In-Fill Deposits (Mixed-Load Fluvial-Tidal Facies Succession F1)
Although it is not always easy to differentiate this faces association from the outer estuary bar cross-bedded sandstone, the cross-bedded sandstones of the innermost parts of the estuary system are distinguished by a greater abundance of erosional surfaces, lag deposits, especially tree trunk logs and vertebrates. This facies is

Table 3.1 Summary of facies and facies associations

Depositional environment	Facies association		Facies
1- Tide dominated estuary	(A) Axial	Tide-influenced fluvial channel (FA1)	Channel in-fill deposits (mixed-load fluvial-tidal Facies Succession (F1)
		Flat laminated sandstone (FA2)	Flat-laminated sandstone (F2)
		Outer estuary bar (FA3) (Tidal channel and tidal bars)	Tabular and trough cross-stratified sandstones (F3)
			Lensoidal calcareous cross-stratified sandstone (F4)
			Sandstone with *Ophiomorpha* burrows (F5)
	(B) Marginal estuary	Tidal flat (FA4)	Inclined heterolithic stratification IHS (F6)
			Heterolithic and Rhythmite beds (F7)
			Mudstone beds (F8)
			Mangroves within cross-bedded sandstone (F9)
2- Open shelf	Bioturbated fossiliferous sandstone (FA5)		Sandstones with moderately to intensively bioturbated branched network *Ophiomorpha* (F10)
			Calcareous, mottled, homogenized, highly bioturbated (*Ophiomorpha*) sandstone (F11)
			Fossiliferous calcareous sandstone, highly bioturbated with *Ophiomorpha* and *Thalassinoides* (F12)
			Fossiliferous ferruginous large scale cross-bedded sandstone with hard crust (F13)
			Glauconitic trough cross-bedded sandstone (F14)
3- Tide dominated delta	Bioturbated fossiliferous carbonate (FA6)		Fossiliferous limestone (F15)
	Coarsning upward (FA7)		Coarse grained Lensoidal heterolithic (F16)
			Thin laminated sand—shale intercalation (F17)
			Homogenized bioturbated sand (F18)
	Fining upward (FA8)		Bioturbated heterolithic sandstone (interbedded mudstone and sandstone) (F19)
			Cross-stratified sandstone with mudstone drapes (F20)

Table 3.2 Description of facies and facies associations

Facies	Characteristics	Interpretation
Channel infill deposits (F1)	Cross-stratified sandstones, with erosional surfaces and lag deposits, especially tree trunks and vertebrates. The logs are bored by the brackish-water borer *Teredolites*	River bars, probably point bars on the innermost sector of estuaries. *Teredolites* suggests that shorelines were close enough during this time to permit brackish water far up into the fluvial system
Flat-laminated sandstones (F2)	Flat-laminated sandstones with some intercalated siltstones (3–4 cm thick)	Upper flow regime Sand Flats in inner estuary
Tabular and trough cross stratified sandstones (F3)	Sets of tabular and trough cross-strata white, well sorted, fine to medium-grained (though coarse grain sizes occur sometimes) sandstones. These sets represents also Alternating thinner/thicker foreset packages (tidal bundles) within subaqueous dunes	Tidally influenced channels
Lensoidal calcareous cross-stratified sandstone (F4)	Large-scale cross-stratified pokilotopic coarse grained sandstones, highly cemented with calcareous cement	
Sandstone with *Ophiomorpha* burrows (F5)	Moderately to intensively bioturbated branched network *Ophiomorpha*	
Inclined heterolithic stratification (F6)	Slump structure–from alternating mudstones and white sandstone layers	Lateral accretion deposits produced by point bars in small muddy tidal channels within the margins of the estuarine system
Heterolithic and Rhythmite beds (F7)	Highly bioturbated Heterolithic and rhythmitic beds. The mudstones are grey with lenticular and wavy cross-lamination	Tidal flat deposits (Marginal tidal dominated estuary)
Mudstone beds (F8)	Laminated intensively bioturbated iron-concretion mudstone beds	Supratidal mudflats occur along tide-dominated coastlines along the margins of estuaries
Mangroves within cross-bedded sandstone (F9)	Cross-bedded fine grained sandstone and root structure	Mangroves colonize upper intertidal (above mean sea level) sediment substrates

(continued)

Table 3.2 (continued)

Facies	Characteristics	Interpretation
Sandstones with moderately to intensively bioturbated branched network *Ophiomorpha* (F10)	Thin laminated, flaser laminated, cross laminated, trough and planar cross stratifications fine grained quatrzose sand, light colour. Bioturbation is moderate to high and includes complex sub horizontal to inclined, branching burrow networks, including *Ophiomorpha*	Open-marine conditions, subtidal environment extended well into adjacent shallow inner shelf environments
Calcareous, mottled, homogenized, highly bioturbated sandstone (F11)	Cross-bedding is mainly trough cross-stratification intensively bioturbated, including *Ophiomorpha* sand with marked grain-size variations, ranging from medium sands, fine to very fine grained, mud drapes are sparse to absent	
Fossiliferous calcareous sandstone, highly bioturbated with *Ophiomorpha* and *Thalassinoides* (F12)	Hard calcareous sandstone, very coarse-grained, fossiliferous with shell fragments from pectin molds and casts, bivalve and *scuttela*. Burrows are *Ophiomorpha and Thalassinoides*	
Fossiliferous ferruginous large scale cross-bedded sandstone with hard crust (F13)	Tangential cross beds that are ebb-current directed, 330–310° coarse to medium grained sand with two hard crusts. The compound cross-stratification on a large scale (2 m thick)	
Glauconitic trough cross-bedded sandstone (F14)	Cross-stratification, trough (350°), the cross lamina 275° glauconitic calcareous sandstone with lag deposits	
Fossiliferous limestone (F15)	Conglomeratic ooidal limestone benthic foraminifera. Other detrital components as feldspar grains (microcline) maybe found as rock fragments. Laterally represent by moderately bioturbated fossiliferous limestone	Open marine environments appropriate for these diverse macrofaunal

(continued)

Table 3.2 (continued)

Facies	Characteristics	Interpretation
Coarse grained Lensoidal heterolithic (F16)	Thinly laminated siltstone and mudstone intercalated with fine grained sandstone. The uppermost part of this bed is represented by 50 cm different sized lenses of coarse-grained quartz	Delta front and prodelta
Thin laminated sand—shale intercalation (F17)	Fissile grey shale with very thin laminated highly ferruginated very fine grain quartzose sand. The shale is more clayey in the base and coarsens to become more silty in the upper part	
Homogenized bioturbated sand (F18)	Quartzose sand, pebbly in places, poorly sorted, homogenized and bioturbated. This facies is shaley in the base and becomes sandier in the upper part	
Bioturbated heterolithic sandstone (F19)	Shale—sand intercalation. Shale is dominant at the base with increasing sand upwards. Moderately bioturbated with gypsum streaks. Ended by coarse grain quartzose	
Cross stratified sandstone with mudstone drapes (F20)	Thin laminated, flazer laminated, tabular and trough cross-stratified fine grained quartz sand, light color. The top most part is cross-bedded with foresets towards 185° Highly bioturbated by *Ophiomorpha* burrows. The top most part is highly cemented, coarse grained, pebbly with abundant mudclasts	Subtidal bars on the delta front reaches of the system

overlain by tabular and trough cross-stratified sandstones without the most proximal signals (tree trunks) (F3) and is underlain by coarse grained lensoidal heterolithic (F21).

Erosional Surfaces
Description
Channelized erosional surfaces of both local extent (simply channel down cutting) and of greater lateral extent (possibly base-level fall incision) are very common in the study succession, and especially here in the most proximal association, these erosional surfaces show relief of up to several meters locally, and in addition the correlation between sites implies 10 s of meters of incision inferred from thickness of the infilling facies. Burrows are recorded below the base of erosional surfaces (section 1, Aux and 2) from ferruginous horizontal *Thalassinoides* (length, 10 cm and diameter 4 cm) and *Ophiomporha*.
Within the sandstone bodies that infill the regional master erosion surfaces additional small erosion surfaces occur and these tend to be lined with clay-ironstone pebbles and abundant chert pebbles of varying dimension and well preserved pieces of petrified wood with long-axis trend 320°, 280°, as well as very large bones mainly oriented along 330°. These east and northeasterly paleocurrents are the ebb-current direction in the estuarine system, as judged by regional knowledge of the direction of the open sea. In the proximal reaches of the system being described here the minor erosion surfaces can sometimes be difficult to distinguish from the master erosional surfaces. In general the cross-stratified sands of the fluvial-tidal reaches have coprolites and low numbers and diversity of bioturbation from *Ophiomorpha* and *Thalassinoid* trace fossils.

Lag Deposits
Description
This facies mantles the master erosional surfaces as well as the minor erosional surfaces within the sand body and consists of different sizes of pebbles and mudclasts, variable in lithology (Fig. 3.1a, Table 3.3).

Tree Trunks
Description
This facies consists of large accumulations of petrified tree logs scattered on the slope, 20–75 cm in diameter and as much as 16 m in length (Table 3.4), and are present in several stratigraphic horizons. The logs are locally bored by the brackish-water borer *Teredolites* (Fig. 3.1b). The trunks lie nearly parallel to each other, with a general northwest-southeast orientation. There are two different types, the first type are the concentric ring trunks and the second type the smooth trunks (Fig. 3.1c). The trunks are surrounded by the cross-stratified sandstones which indicate the same paleoflow direction as the trunks themselves (Fig. 3.1d).
Rose diagram constructed for the frequency of silicified trunks direction and indicates four major sets (Fig. 3.2a, WP. 203), five major sets (Fig. 3.2b, WP. 202) and seven major sets (Fig. 3.3a, WP. 206, b, WP. 217).

Fig. 3.1 Photographs of common lithofacies in tide-influenced fluvial channel deposits (FA1). **a** Common lithofacies in the lower part of the upward-fining packages that include lag deposits, the *arrows* outline a sharp erosional contact at the base of an upward-fining package. **b** Outcrop scale photograph of typical the logs are locally bored by the brackish-water borer Teredolites **c** In-situ fossilized tree showing the main trunk (*T*), smooth trunks. **d** The trunks covered by the cross-bedded sandstone in the same direction as the trunks

Vertebrates

Description

Vertebrate fossil fragments were recorded along the erosion surfaces of this facies association. The recorded vertebrates are listed below:

Turtles, two isolated Anthracothere premolars, two *Sivameryx moneyi* molars, small giraffid cf. Paleomerycid, crocodile tooth, *Sivameryx moneyi dental fragments, Sivameryx moneyi molar, Brachyodus depereti* molar, *Gomphothere* tooth fragments (Fig. 3.4a), reptile vertebra, *Brachyodus depereti* upper molar, MISC. isolated shark and crocodile teeth, parts of Anthracothere jaw cf. *Brachyrdus depereti*, fragment, premolar region *Brachyrdus depereti*, fragment, premolar region *Sivameryx moneyi*, MISC. isolated teeth Anthracothere, shark, crab parts and crocodile small creodont, isolated pig tooth cf. *Xenochoerus*, molar fragment cf. *Brachyrdus depereti*, Anthracothere, MISC. teeth and bones, *anthracothere* tooth, crocodile, ray, shark, fish vertebrate, jaw fragment *Brachyrdus depereti*, giraffid molar, *Doratherium* lower molar, small carnivore jaw teeth, *Anthracothere* tooth, small canine, possible primate, *Anthracothere* fragments, *Gomphothere* tooth fragments, 2 crab parts, MISC. bones and teeth, Artiodactyl astragalus,

Table 3.3 Lag deposits section 7

Section no. 7	Length (cm)	Width (cm)	Short (cm)	Composition
Unit II	5.5	4	2	Mudstone
	8	5	2	Mudstone
	6.5	4	1	Mudstone
	4.5	4	1	Mudstone
	3.5	2.5	1	Mudstone
	4.5	3.5	1	Flint
	4.5	3.5	0.5	Flint
	5.5	1.5	1.5	Mudstone
	5	3	1.5	Mudstone
	1.5	4	0.5	Mudstone
	1.5	1	0.2	Flint
	1.5	0.5	1	Flint
	2.5	1.5	1.2	Flint
	2.5	2	1.4	Flint
	2.5	1.8	1	Flint
Unit IV	3.2	2.2	0.9	Ferruginous sandclast
	3.2	3.2	1.1	Ferruginous sandclast
	1.4	1	0.6	Flint
	2	0.8	1.5	Ferruginous sandclast
	3.8	1.5	0.9	Flint
	2.4	2	1	Flint

MISC. teeth and bones (turtle, shark, anthracothere premolar) upper (L) maxillary *Prolibytherium maqnuri* w/P4, M1-3 and (R) premolar, mandible *Sivameryx moneyi*, upper left maxillary teeth cf. *Afromeryx africanus*, unidentified mandible fragment, small mammal proximal ulna (L), Artiodactyl astragalus, Big mammal toe, MISC. bones and teeth crocodile tooth, large reptile vertebra, fish vertebra, mammal toe, MISC. bones and teeth, includes small mammal radius, Anthracothere upper molar, may be bird femur, dP4 *Xenochoerus africanus*, 2 isolated teeth, includes 1 inciser and 1 fish tooth, Palate of *Xenochoerus africanus* juvenile, (R) P2-M3, (L) P2-M3, *Anthracothere* lower molar, small palate with tooth roots, MISC. bones, Coprolite, Fragment Doratherium jaw, fragment of a crocodile jaw MISC. bones and teeth, MISC. bones, Anthracothere skull cf. *Sivameryx maneyi* Anthracothere metapodial cf. *Brachyrdus depereti*, *Anthracothere* mandible *Brachyrdus depereti*, MISC. fish and reptile bones, MISC. fish vertebra and teeth MISC. fish and reptile, *Turatella*, ray, shark, MISC. bone, MISC. bones and teeth, large vertebra, large mammal calcaneus, Artiodactyl astragalus, 2 fused radius - ulna large mammal, fragment of Rhinocerotid tooth, large carnivore jaw, crocodile jaw fragments, mammal vertebra, large carnivore jaw, skull of *Brachyodus depereti* (Fig. 3.4b), ungulate metapodial, *Brachyodus depereti* astragalus, *Canthumaryx sirtensis* horncore, small reptile vertebra, (R) maxilla *Sivameryx moneyi* P3-M3, (R) mand (Fig. 3.4c). Fragment with M3 Brachyodus depereti, small mammal proximal ulna, small artiodactyl talus cf. *Dorcatherium*,

Table 3.4 Paleodirections of silicified wood trunks and their lengths

WP	Directions (°)	Length (m)	Description
202	355	8.5	
	352	3	
	345	1.5	
	347	7	
	350	3	
	310	8	
	255	8	
	210	4	
	205	8	
	185	3	
	320	5	
	320	3.5	
	350	4	
	347	8.5	
203	350	8	1-Unit 6
	355	2.5	2-Large trunks not similar to unit or unit 10
	357	16	
	335	9	
	315	8	
	333	5	
	352	3	
	345	3.5	
	355	55	
	342	20	
206	355	1	1-No palm found, abundance to concentric rings
	310	2.5	2-May be unit 8
	275	2	
	255	1	
	235	1.5	
	220	2	
	330	3	
	265	1	
217	303	11	
	275	15	
	350	7	
	285	3	
	315	14	
	235	14	
	285	4	
	290	8	
	33	13	

Brachyodus depereti (R) max. M1-2, *Sivameryx moneyi (R)* max P4- M2, 2 ulnae: 1 large, 1 small, medium sized artiodactyl astragalus, *Xenochoerus africanus juv. Mand. 1/2 dp4, M1, M, Brachyodus depereti* astragalus, *Sivameryx moneyi juv.*

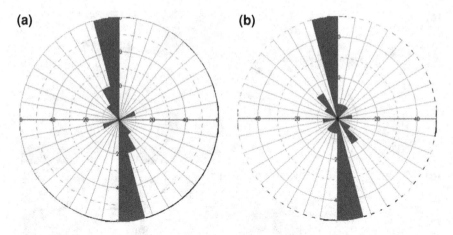

Fig. 3.2 Rose diagram representing the main trend directions of the trunks. **a** WP 203. **b** WP: 202

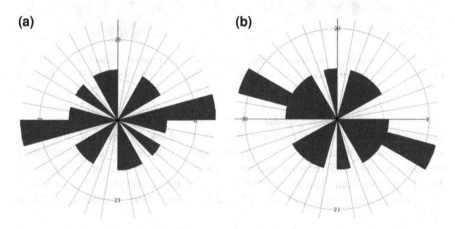

Fig. 3.3 Rose diagram representing the main trend directions of the trunks. **a** WP 206. **b** WP: 217

Mand. P4-M2, *Sivameryx moneyi* fragments mand. *M3*, 2 bird bones 2 worn Sivameryx moneyi mandibles, *Brachyodus depereti (L) mand. P2–P3, P magnieri mand.* (L) M1-3, mammal distal femur, *Brachyodus depereti (L) M3, Canthumaryx sirtensis* horncore, Anthracotherce skull cap (2 Pieces), *Dorcatherium tooth,* caudal vertebra, small mammal tibia epiphysis, Reptile vertebra, Garial snout, snub nosed crocodile snout, *P. magnieri* jaw (L) mand. P4- M3, *Sivameryx moneyi* (L) M2-3, anthracothere mand. Very worn, mammal tibia, fragment probosciden tooth, *Canthumaryx horncore* (Fig. 3.4d), very worn anthracothere jaw, *Dorcatherium* (L) max. P2-M3, worn anthracotherce lower molar.

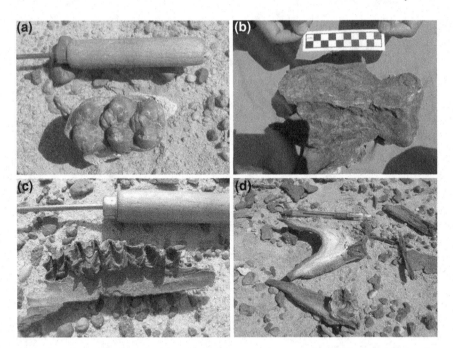

Fig. 3.4 Photographs of common vertebrates in tide-influenced fluvial channel deposits (FA1).
a *Gomphothere* tooth. **b** Partial *Brachyodus* skull. **c** *Sivameryx moneyi* (*R*) max. **d** *Canthumeryx*
horn-core

Interpretation of the channel infill

Stacked sets of cross-strata, erosional surfaces, basal lag deposits, and fining- and
thinning-upward trends strongly suggest that this facies, which immediately
overlies each regional erosion surfaces, represents a complex of amalgamated
fluvial-tidal and tidal-channel and bar deposits in the proximal reaches of a tide-
dominated estuary or estuarine bay (Johnson and Levell 1995; Wonham and Elliott
1996).

Over the years, the majority opinion has held that the tree-trunk facies represents fluviatile facies. However, tidal influence and marine affinity is quite clear for
the trunks and should not be overlooked. The marine affinity tends even to
overshadow the fluviatile affinity in places. The geologists who have worked in the
area commonly regard this facies as a fluviatile facies (Said 1962), Stratigraphic
Subcommittee of the National Committee of Geological Sciences (1974), reported
by El Gezeery and Marzouk (1974). But the presence of *Teredolites* suggests that
shorelines were close enough during this time to permit marine and freshwater
mixing far up into the fluvial systems. The *Teredolites* is considered as the most
diagnostic indicator to the marine/brackish water trace fossil (Roberts 2007).

The question arises as to how the trunks originated and came to their present
position. Two possibilities have been suggested. The trunks come downriver and
as they floated in the water channels became water-logged and became heavier so

as to sink to the base of the channel. The cross-strata in the sand deposits take the same direction as the trunks. Alternatively, the trunks became oriented in direction perpendicular to the water flow direction. The first idea relates to their occurrence near from the channel axis and the second idea to their occurrence farther downstream in the estuary. This problem can be solved temporary by looking at the orientation of the trunks. In addition, it needs paleo-botany investigation in the near future to solve it.

What is the Origin of the Petrified Wood?

Silicified wood occurs in these deposits, because of the presence of dissolved silica within the groundwater. The silica is derived from the dissolution of the volcanic material by the groundwater within the volcanics or sediments. This dissolved silica in the form of monomeric silicic acid attaches itself to the lignin and cellulose of the wood. With time, a layer of the monomeric silicic acid forms on the exposed woody tissues. The monomeric silicic acid dehydrates into silica gel. Additional layers of the monomeric silicic acid attach to this silica gel eventually filling and encasing the wood with silica gel. A rapid loss of water converts the silica gel into amorphous silica (opal) (Leo and Barghoorn 1976; Scurfield and Segnit 1984).

Within 10–40 million years, the opal of the silicified wood further dehydrates and crystallizes into microcrystalline quartz (chert). Factors such as pressure and temperature may speed or slow the process, but eventually the opal of the silicified wood becomes chert (Stein 1982). During the change from opal to chert in silicified wood, the relict woody texture may either be retained or lost.

Mid-Estuary Flat-Laminated Sandstone (FA2)

Description

The facies consists of 120 cm (bed 1-1-4, unit I, section 1), thin laminated intercalated sand and silt (Fig. 3.5). The thickness of laminae is 1–2 cm with sharp contact with each other. This bed is only slightly burrowed. The sand is coarse grained and the silt is arranged into bundles with 3–4 cm thick. This facies is

Fig. 3.5 Photographs of common lithofacies in mid-estuary flat-laminated sandstone (FA2). Outcrop scale photograph slightly burrowed thin laminated intercalated sand and silt

overlain by thin laminated sand-shale intercalation (F22) and is underlain by calcareous, mottled, homogenized, highly bioturbated (*Ophiomorpha*) sandstone (F12).

Interpretation

The plane-parallel lamination in the fine-grained sandstones and the lack of interlaminated mudstones suggest upper flow-regime deposition. The plane-parallel laminated beds formed at maximum tidal current energy (Kreisa and Moila 1986). This facies is typical for upper-flow-regime sand flats in the inner to middle parts of tide-dominated estuaries and occur seaward of the mouths of fluvial-tidal channels, in the narrow, landward end of the estuarine funnel, where the tidal current speeds are highest (Dalrymple 1992; Dalrymple et al. 1992; Wells 1995).

Outer Estuary Tidal Bars and Channels Association (FA3)
Tabular and Trough Cross-Stratified Sandstones (F3)

Description

Thickly stacked sets of cross-stratified sandstones and it is one of the most common facies in the study succession, extend over lateral distances of hundreds of meters, and vary considerably in thickness (25 m, unit X, 3.5 m, unit VIII, 8.5 m unit VI, section 5, 53 m, unit VIII, section 6, 20 m, unit VIII, section 4, 17 m, unit VI, section 7, 14 m, unit II, section 1). This facies is consisting of white, well sorted, fine to medium-grained (though coarse grain sizes occur sometimes) sandstones. The cross-bedded sets vary in grain size fine to medium-grained and the sets averaging 0.1–0.9 m thick. Both grain size and set thickness typically decrease upwards, giving rise to fining and thinning upward units that attain a thickness of 1–1.5 m. Some channel fills show compound cross-strata characterized by medium to large scale (0.3–0.9 m thick), low-angle dipping (5°, 10° and 40–45°) cross sets with superimposed smaller scale (0.1–0.2 m thick) sets of tabular and trough cross-strata. Paleocurrent data measured from cross-sets (including compound sets) show a main NNW (340°, 330°/40°, 325°/10°, 324°/5°) orientation but others are showing 140°, (section 21) orientation. This south-easterly direction is likely to be the flood-tide direction into the estuary, and such flood tidal signals would be expected to become stronger towards the outer reaches of the estuary. In addition, the cross set boundaries are either straight or undulating and demarcated by mud drapes. Mudstone rip-up clasts (1.5 and 11–14 cm) are frequent (Fig. 3.6a), either in concentrates at set boundaries or dispersed within sets. A characteristic feature of this facies is the presence of lateral successions of alternating thinner/thicker foreset packages (tidal bundles) or mud drapes that form couplets (Fig. 3.6b). The reactivation surface could be found but we can't trace it accurately.

Channel deposits of this type are volumetrically very important and constitute approximately 70 % of the Moghra successions. This facies is overlain by sandstone with moderately to intensively bioturbated branched network *Ophiomorpha* (F10) and underlain by channel in-fill deposits (F1).

Fig. 3.6 Photographs of common lithofacies in outer estuary tidal bars and channels Association (FA3). **a** Mudstone rip-up clasts dispersed at the base of this facies. **b** Outcrop scale photograph of typical planar cross-stratified sandstone (Sp) with alternating thinner/thicker foreset packages (tidal bundles) or mud drapes that form couplets (*yellow arrows*)

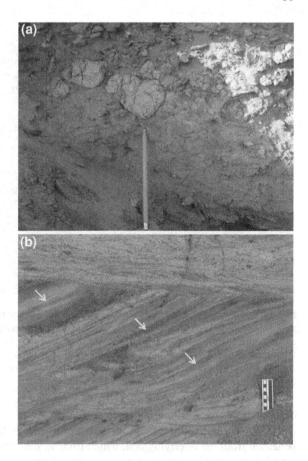

Lensoidal Calcareous Cross-Stratified Sandstone (F4)

Description

This facies consists of large-scale cross-stratified sandstones, highly cemented with calcareous cement. This facies is showing poikilotopic coarse grained planar cross-stratified sandstone beds (Fig. 3.7a, b) are commonly less than 1 m thick, sometimes this bed discuttered of pebbles on individual foresets (Fig. 3.7c). Note the lack of bioturbation in this lithofacies. However, this facies is restricted in its extension, it is found in most sand-body units close to the erosional surfaces. This facies is overlain by tabular and trough cross stratified sandstone (F3), and underlain by channel in-fill deposits (F1).

Ophiomorpha Burrows (F5)

Description

This facies is restricted to section 20 (top of bed no. 15, unit XIV and top of bed no. 10, unit X), section 22 (bed no. 14, unit XII) and section 25 (bed no. 7, unit XIII). This facies is represented by vertical, oblique, and horizontal *Ophiomorpha* burrows. This trace fossil is similar to the *Ophiomorpha nodosa* which it is

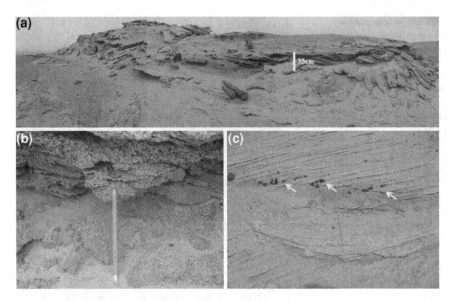

Fig. 3.7 Photographs of common lithofacies in outer estuary tidal bars and channels facies association (FA3). **a** Panoramic view of the outcrop showing poikilotopic coarse grained planar cross-stratified (Sp) sandstone beds are commonly less than 1 m (3 ft) thick, the most common lithofacies in facies. **b** Close-up of poikilotopic calcitic sandstone grains. **c** Close-up of planar cross-stratified sandstone (Sp) with pebbles on individual foresets (*arrows*). Note the lack of bioturbation in this lithofacies

described as of a three-dimensional network of coarsely pelleted burrows (Fig. 3.8a–d). In addition, these burrows are lined by clay. Most of these burrows penetrate from the overlying bed, which truncates the upper part of the tabular and trough cross stratified sandstones (F3). This facies is overlain by sandstone with moderately to intensively bioturbated branched network *Ophiomorpha* (F10) and underlain by the former facies F3.

Interpretation (F3–F5)

Bidirectional cross-strata and foresets with mud drapes, foreset bundling and mud couplets indicate deposition in tidally influenced channels. Elliott (1986) considered cross-bedded sandstone with an erosional base, basal lags, and fining-upward trends to represent estuarine tidal channels (see also Shanley et al. 1992; Shanmugam et al. 2000).

The presence of bidirectional paleocurrents, double mud drapes, rip-up clasts and a marine ichnofacies suggest a tidal origin for these channels (Olsen et al. 1999). Double mud drapes along sets and foresets are diagnostic of the systematic slack-water periods characteristic of a tidal environment (Boersma and Terwindt 1981; Smith 1988; Visser 1980), and the abundance of large-scale cross-beds with alternating opposite paleocurrent directions indicates strong bipolar currents, as would be expected from ebb and flood currents that alternate in tidal channel systems (De Raaf and Boersma 1971; Reineck and Singh 1973; Yoshida et al. 2004).

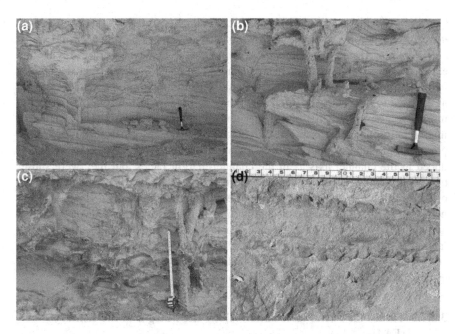

Fig. 3.8 Photographs of common trace fossils in outer estuary tidal bars and channels facies association (FA3). **a** *Vertical*, *oblique*, and *horizontal* burrows occur in the upper part of this facies association, including *Ophiomorpha nodosa* burrows. (**b**, **c**) Two sizes of *Ophiomorpha*. Pellets are less uniformly packed and the specimens of *Ophiomorpha* nodosa preserved in full relief. **d** Close view to the plan view of this trace fossil

Facies F3-5 therefore records the central area of tidal channel at or near the valley axis. At such locations flow energy was high enough to transport sand as dunes, compound dunes and bars (Rossetti and Júnior 2004). The compound cross-sets record the highest volume of sand accumulation in the estuarine channels and attest to the presence of medium-and small scale bed forms superimposed on low angle dipping, large scale bedforms.

In addition, *Ophiomorpha* is one of the best-known trace fossils for paleontologists and sedimentary geologists due to its abundance in Mesozoic and Cenozoic shallow and marginal marine deposits. The ichnogenus designates multiple-branching gallery systems of variable complexity characterized by having a thick pelletal lining (De Gibert et al. 2006). It is a softground-related ichnofacies that generally indicates active sediment accumulation (low to high rates) on moist to fully subaqueous depositional surfaces, and hence is associated with conformable successions. All softground ichnofacies like *Ophiomporpha* are associated with presence of water and active sediment aggradation. Moreover, the trace fossil *Ophiomorpha*, particularly the ichnospecies *O. nodosa*, is widespread and abundant in shallow marine sands and marginal marine of Cretaceous to Pleistocene age exposed along the Atlantic and Gulf Coastal Plains (e.g., Curran 1985; Curran and Frey 1977; Erickson and Sanders 1991; Frey et al. 1978; Martino and

Curran 1990; Pickett et al. 1971). At the Pollack Farm Site, specimens of *O. nodosa*, that closely resemble burrows of the modern callianassid shrimp *Callichirus major* (formerly *Callianassa major*), and associated trace fossils are exposed in tidal or estuarine channel sands. The results indicate that the *Ophiomorpha*-producers, presumably callianassid shrimp, preferred the channel margin as opposed to the center of the channel, may be because of lower current velocity, more stable substrates for burrowing, and other related factors. The alternative interpretation, that the observed difference in density results from differences in preservation, is not supported. In this scenario, as many *Ophiomorpha* were produced in the channel axis as in the channel-margin facies, but they were subsequently eroded. Because callianassids burrow deeply (>50 cm), complete removal of their burrows by erosion would require multiple large erosional events that left no record (e.g., major scours, discontinuities) in the sedimentary sequence and that are not consistent with the in-channel accretion reflected by the trough-cross-laminated sands (Miller et al. 1998). In brief, all of the aspects of these facies strongly suggest that they are corresponding to the outer zone of tide-dominated estuary model of Dalrymple et al. (1992).

3.2.1.2 Tidal Flat Facies Association Along the Margin of the Estuary (FA4)

Inclined Heterolithic Stratification IHS (F6)
Description
This facies is variable in thickness (6 m, unit 15, section 20 and 1.5 + 1.2, unit IV, section aux) and consists of alternating mudstones and white sandstone layers. The massive grey mudstones, are highly brecciated. The muds are unbioturbated and unlaminated. Sandstones are very fine grained; sometimes contain ripple laminae. This IHS facies displays low-angle dipping foresets that downlap onto a large erosional surface and the deposits have internal erosion surfaces (Fig. 3.9a). Soft-sediment deformation structure (slump structure) is well-developed particularly in the uppermost parts of certain interval (Fig. 3.9b). This facies is overlain by fossiliferous calcareous sandstone, highly bioturbated with *Ophiomorpha* and *Thalassinoid* (F13) and is underlain by sandstone with moderately to intensively bioturbated branched network *Ophiomorpha* (F10).

Interpretation
The well-developed inclined heterolithic strata are interpreted as lateral accretion deposits produced by point bars in small muddy tidal channels within the margins of the estuarine system (Gingras et al. 1999; Shanley et al. 1992; Thomas et al. 1987) (e.g., Thomas et al. 1987; Shanley et al. 1992; Gingras et al. 1999 cit in Finzel et al. 2009). The term "inclined heterolithic stratification" was formalized by Thomas et al. (1987) and he mentioned that these heterolithic intervals display an original depositional dip for the orientation of sandstone and mudstone beds, interpreted to record the depositional dip of tidally influenced (tidal-fluvial) point bars (MacEachern and Bann 2008). Semantically, the term is purely descriptive and

Fig. 3.9 Photographs of
common lithofacies in tidal
flat Facies Association along
the margin of the estuary
(FA4). **a** Outcrop scale
photograph of inclined
heterolithic and lateral
accretion surfaces. Note the
several channel-base and
internal erosion surfaces.
b Small-scale slumps and
soft-sediment-deformed
mudstone (yellow pencil for
scale)

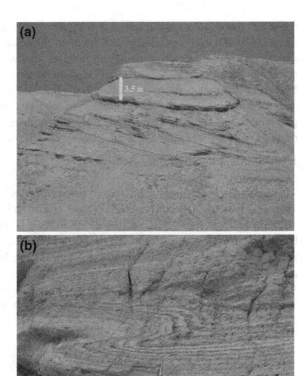

applied to concordant sets of inclined beds that display pronounced lithological
heterogeneity. In addition, Thomas et al. (1987) diligently note the various depositional settings in which IHS might occur. Prominent among these are tidal settings,
in particular on tidally influenced point bars. Since that publication, IHS has been
associated increasingly with tidally influenced, marginal marine deposits (Smith,
1989; Ranger and Pemberton, 1992; Räsänen et al. 1995; Falcon-Lang 1998; Witzke
et al. 1999; Gingras et al. 1999). In general, due to the common presence of IHS in
tidal settings, it is intuitive that the inter-beds result from diurnal or semidiurnal tidal
rhythmicity. However, even with cm-scale interbeds, the perceived sedimentation
rates would be anomalously high. Furthermore, IHS beds deposited in brackish
water settings commonly preserve trace fossils, suggesting that in those instances
sedimentation rates are not extraordinary. Also, some authors have suggested that
marginal marine IHS simply reflect the shifting of the turbidity maximum due to
fluctuations of continental discharge (Gingras et al. 1999; Smith 1989).

The origin of the massive grey mudstone would be a depositional event of
fluidized mud that results in centimeter- to decimeter-thick mudstone beds
(the development of fluid-mud deposits (i.e., structureless mud layers more than

0.5–1 cm thick that were deposited in a single slack-water period, Dalrymple and Choi 2007). These types of deposits occur in the middle part of modern estuary systems where sediments carried from both the landward and seaward directions converge, resulting in an area of high-suspended sediment concentrations. Mud- and clay-size particles flocculate in the brackish water to form heavier grains and facilitate deposition. These fluid-mud deposits commonly occur in topographically low areas, in channel bottoms for example, and are therefore bounded by beds of cross-stratified sandstone (Finzel et al. 2009).

Heterolithic and Rhythmite Beds (F7)

Description

This facies occurs as highly bioturbated heterolithic and rhythmitic beds that occur sporadically in the studied succession (Fig. 3.10a, b). It is variable in thickness from 10 m (beds 25-11, unit, XIII, section 25), 8 m (unit I, section 8) and characterized by an alternation of sandstone and mudstone. Sandstones are egg-yellow, fine-grained and slightly to highly bioturbated (Fig. 3.10c). Mud drapes are widespread along the foresets of cross-stratification. The mudstones are grey with lenticular and wavy cross-lamination and sometimes contain gypsum streaks. Sparse pebbles occur at the base of some of the mudstones. In section 8 the sand-shale intercalation

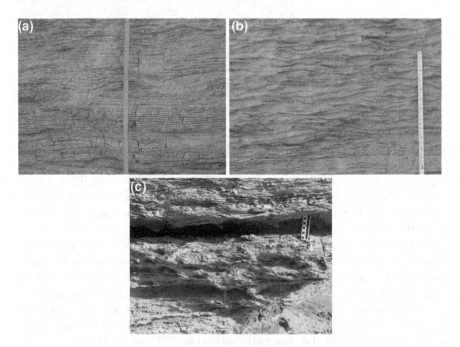

Fig. 3.10 Photographs of common lithofacies in tidal flat facies association along the margin of the estuary (FA4). **a** Outcrop scale photograph of heterolithic, wavy bedded unit, with oscillation-rippled sandstone bed draped with mudstones. **b** Large scale mud draping lamina in heterolithic bed. **c** Bioturbated heterolithic, wavy bedded

is variably in thickness thick. Sand layer thickness varies between 20 and 30 cm whereas the mud layers are about 5 cm. The upper part of this facies in section 8 grades upward into 4.5 m thick mudstone with highly ferruginous fine-grained sandstone. This facies is overlain by cross-stratified sandstone with mudstone drapes (F25) of the fining up-ward (FA8) and underlain by fossiliferous calcareous sand-stone, highly bioturbated with *Ophiomporpha* and *Thalassinoid* (F13).

Interpretation
Tide-dominated estuary margins preserve muddy and heterolithic intervals inter-preted as tidal-flat deposits (MacEachern and Bann 2008). The marine trace fossils, such as *Ophiomorpha*, suggest a tidal origin for these deposits (Olsen et al. 1999). The abundant but low-diversity ichnofauna assemblage indicates a highly stressed environment and is consistent with an estuarine setting in which shifting substrate and fluctuations in salinity, temperature and sedimentation rates are common (Clifton 1983; Pemberton 1992; Smith 1988). The abundant flaser and lenticular bedding are features common in tidal flats (Allen 1991; Dalrymple 1992; Middleton 1991).

Mudstone Beds (F8)
Description
This facies occurs as intensively bioturbated iron-concretion laminated mudstone beds (Fig. 3.11a). The mudstone is grey in color and variable in thickness (1.5 m, unit I, section 1, 1 m, unit 1, section 2 and 2 m, unit IX, section 20). The bioturbation varies from slightly to intensively burrows and represented by *Ophiomorpha* trace fossil. Base of this facies unexposed and the topmost part is characterized by gradational contact. This facies is overlain by channel in-fill deposits (F1) and underlain by fossiliferous calcareous sandstone, highly biotur-bated with *Ophiomorpha* and *Thalassinoid* (F12). This facies in section 7 is rep-resented by 2 m massive dissected mudstone with planet remains. This facies when it is represented by plant remains (For example section 7) it is overlain by channel in-fill deposits (F1) and underlain by tabular and trough cross stratified sandstone (F3).

Interpretation (SF1 and SF2)
The organic-rich mudstone with paleosols indicates occasional deposition by suspension fallout in a quiet-water environment, close to land in areas that were regularly subaerially exposed. This Facies is interpreted as a supratidal mudflat (see also Dalrymple et al. 1992). Supratidal mudflats occur along tide-dominated coastlines, along coastlines sheltered by barrier islands, and along the margins of estuaries (Pontén and Plink-Björklund 2009). The occurrence of plant fragments suggests a nearshore location (Hampson et al. 2008).

Mangroves Within Cross-Bedded Sandstone (F9)
Description
This facies is restricted to section 5' and is about 7 m thick (bed 5'-5-1). It consists of cross-bedded, fine grained sandstone rich in root structures (probably man-groves) (Fig. 3.11).

Fig. 3.11 Photographs of common lithofacies in tidal flat facies association along the margin of the estuary (FA4). **a** Outcrop scale photograph of bioturbated (*Ophiomorpha* burrow with *yellow arrow*) iron-concretion laminated mudstone beds. **b** Mangroves structure

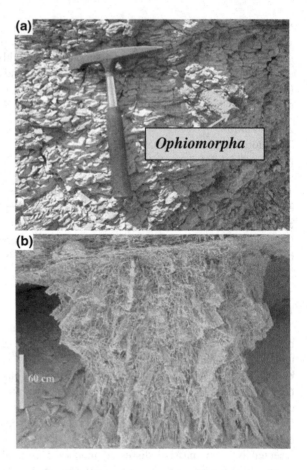

This facies is overlain by mudstone bed (F8) and the base of this facies in this section (5′) is unexposed but could be this facies is underlain in the equivalent section no. 5 by tabular and trough cross stratified sandstones (F3) and *Ophiomorpha* burrows (F5) or could be calcareous, mottled, homogenized highly bioturbated (*Ophiomorpha*) sandstone (F12).

Interpretation

Given the present widespread occurrence of mangroves, their apparent long evolutionary history, and their influence on intertidal sedimentology, it seems reasonable to suggest that mangroves have played an important role in intertidal sedimentology for long periods of geologic time (Perry et al. 2008).

Mangroves are plants (mostly tree species) that are physiologically adapted to live in saline conditions, and can densely vegetate the intertidal areas of tropical coastlines (Walsha and Nittrouer 2004). Several adaptations, including prop roots, pneumatophores, viviparous seedlings (propagules), and salt exclusion/secretion methods, allow mangroves to survive in a wide range of salinities (0–90 %) and in anoxic sediments (Clough 1992; Clough et al. 1992; Saenger 1992).

Mangrove environments are associations of salt tolerant trees, shrubs, palms, and ferns that colonize upper intertidal (above mean sea level) sediment substrates. They presently cover around 190,000 km^2 (Spalding et al. 1997), and although most diverse and extensive in the humid tropics, they extend into warm temperate climatic zones and occupy a broad spectrum of coastal environments including those in desert and semi arid areas, and on offshore islands (Duke 2006; Plaziat 1995; Woodroffe 1992). Their current distribution can be reasonably correlated within the equatorial belt by the mean 15 °C sea-surface temperature isotherm (Woodroffe and Grindrod 1991), which approximates to a latitudinal range between 30°N and 30°S. McGill (1958) estimated their range to include 60–75 % of tropical and warm temperate coastlines. Despite their extent and diversity in modern coastal settings, the mangrove fossil record is far from complete, but mangroves appear to have a long evolutionary history (Perry et al. 2008). Mangrove environments are particularly interesting from a sedimentological perspective because they: (1) strongly influence intertidal sediment accumulation within both transgressive and regressive phases of coastline evolution (e.g., Carter et al. 1993); (2) have considerable application in paleoecological reconstructions as proxies for ancient shorelines and as indicators of climatic zones (e.g., Plaziat 1974); and (3) are important sites of sediment accumulation and organic matter production. For example, mangrove roots may baffle currents and enhance sediment settling and stabilization (Ellison 1988; Furukawa and Wolanski 1996; Furukawa et al. 1997; Woodroffe 1992). In areas of high sediment supply and accumulation, rapid shoreline (and mangrove) progradation may occur (Allison et al. 2003), whilst below-ground organic-matter production may promote rapid vertical substrate elevation (Rogers et al. 2005). Above ground organic-matter production rates are also often high, making significant contributions to intertidal substrates (Woodroffe et al. 1988). Mangrove ecosystems strongly influence the geomorphology and sedimentology of contemporary intertidal environments and are known to have occupied similar ecological niches at least as far back as the early Cenozoic (Perry et al. 2008).

The roots of these sea-margin trees became concentrically encrusted in carbonate, probably in part due to seasonal climate. Calcareous rhizoconcretions of this stump include upward-branching pneumatophores indicating that it was an ancient mangrove. The roots of this intertidal tree have become concentrically encrusted with carbonate, probably while the tree was alive. The three-dimensional exposure of these rhizoconeretions is due to wind erosion in this barren desert region (Bown and Kraus 1988; in Retallack 1977).

3.2.2 Open Shelf

Bioturbated-Fossiliferous Sandstone Facies Association (FA5)

"Highly bioturbated" indicates that physical sedimentary structures and trace fossils are rarely preserved; 'moderately bioturbated" means that zones characterized by physical sedimentary structures predominate over burrowed intervals and that identifiable trace fossils are common (Martino and Curran 1990).

Sandstones with Moderately to Intensively Bioturbated Branched Network *Ophiomorpha* (F10)

Description

This facies comprises of (6 m, bed no. 8-1-1, unit I, section 8) thin laminated, flaser laminated, cross laminated, trough and planar cross-stratifications mainly fine grained quatrzose sand, light color with highly ferruginated thin laminae at the base. Bioturbation is moderate to high and includes complex sub horizontal to inclined, branching burrow networks, including *Ophiomorpha*. *Ophiomorpha* is characterized by branching, three-dimensional burrow systems that comprise shafts and tunnels containing well developed linings of agglutinated sediment. These extensive burrows are commonly lined with iron oxide cement. Burrow linings are smooth on the interior and they are characteristically pelleted on the exterior (Figs. 3.8b–d and 3.12a). Burrow systems range from simple to complex, with irregularly spaced, Y-shaped branches. The burrow walls are 1–2 mm thick with smooth interiors and mammalated exteriors. The poorly formed pellets are approximately 2 mm in diameter. Some burrows have unequal pellet distribution, and isolated parts of the burrow system have smooth, thin walls; this is common for *Ophiomorpha*-Y-shaped branches are found in the *Ophiomorpha* bed.

Cross-beds have foreset mud drapes and, less commonly, tidal bundles. This facies occurs in the lowest 1.5 m thick (bed no, 1-4-1, section 1) massive, highly bioturbated yellowish color, fine grained quartz sand probably calcareous rich in mud clasts, the clasts mainly shale fragment, concretion and sandstone clasts. Concretion coated with iron oxides, the bed is hard and consolidated ledge forming. In section 6 (unit VI) this facies is consists of 1 m coarse grain, gravelly, sandy characterized by large-scale cross bedding with very characteristic burrows, honeycomb shape in the plane view and cylindrical in vertical (Fig. 3.12b).

This facies association is overlain by calcareous, mottled, homogenized, highly bioturbated (*Ophiomorpha*) sandstone (F12) and underlain by the tabular and trough cross stratified sandstone (F3) and *Ophiomorpha* burrows (F5).

Calcareous, Mottled, Homogenized, Highly Bioturbated (*Ophiomorpha*) Sandstone (F11)

Description

This facies consists of (1 m, unit 1, section 4, 1.6 m unit 1, section 21, 2.5 m, unit III, section 6 and 3 m, unit XI, section 22) sand with marked grain-size variations, ranging from medium sands, fine to very fine grained. The mud drapes are sparse to absent. The bioturbated sandstone units are, fine to very fine grained and are intensively bioturbated, including *Ophiomorpha*. This facies is characterized by pale brown to egg yellow color, ferruginated, calcareous, mottled, homogenized, highly bioturbated (*Ophiomorpha*) sandstone (Fig. 3.12c, d). Sometimes the burrows are recognized only by the grey color filling material or with clay-lined burrows. This facies in overlain by sandstone with moderately to intensively bioturbated branched network *Ophiomorpha* (F10) and is underlain by tabular and trough cross stratified sandstone (F3).

Fig. 3.12 Photographs of common lithofacies in bioturbated–fossiliferous sandstone facies association (FA5). **a** *Ophiomorpha*. **b** Burrows, honey-comb shape in the plan view and cylindrical in *vertical*. **c, d** Mottled, homogenized and highly bioturbated (*Ophiomorpha*) sandstone

Fossiliferous Calcareous Sandstone, Highly Bioturbated with *Ophiomorpha* and *Thalassinoides* (F12)

Description

This facies (bed 20-11, unit XI, section 20) consists of 1 m thick hard calcareous sandstone, very coarse-grained, fossiliferous with shell fragments from pectin molds and casts, bivalve and *scuttela* (Fig. 3.13a, b). The cross bedding is mainly trough cross-stratification (Fig. 3.13c). Burrows of different sizes and direction, vertical and inclined are recorded. Some of these are identified as *Ophiomorpha* in the top of this facies and *Thalassinoides* also present (Fig. 3.13d). It is also characterized by tabular cross bed (220°). *Thalassinoides* is characterized by three-dimensional burrow systems that comprise smooth-walled, cylindrical segments with Y- to T- shaped branching. The unlined burrows are approximately 1.0–2.0 cm in diameter. The larger diameter is associated with beds of higher diversity. This facies can also represented in (section 20, 1.6 m, unit IX) and consists of marked grain size variations, ranging from coarse to medium grained sands, sparse pebbles along this facies. Sands are egg yellow calcareous sandstone with fossils fragments from pelecypods, echinoid fragments and some gastropods molds and casts (Fig. 3.14a, b). This fossiliferous sandstone is moderately to intensively bioturbated (Fig. 3.14c). Sometimes this subfacies association is

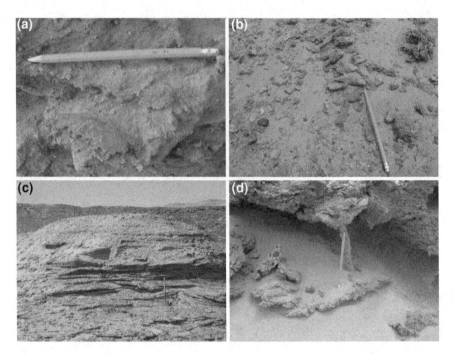

Fig. 3.13 Photographs of common lithofacies in bioturbated-fossiliferous sandstone facies association (FA5). **a** Shell fragment from *Scutella* within bioturbated sandstone. **b** Plan view of shelly (*Turritella* dominated) fossiliferous sandstone. **c** Trough cross-stratification calcareous Sandstone. **d** *Thalassinoides* trace fossil

phosphatic and cavernous. The external geometry forms resistant ledges. This facies is overlain by channel in fill deposits (F1) and underlain by tabular and trough cross stratified sandstones (F3).

Petrography this facies is represented by sidertize bioclastic bearing quartz arenites (Fig. 3.15a–d) which it consists mainly of quartz grain 50–60 % or more (Fig. 3.15a, b), subrounded to rounded, spherical and elongated grains. The majority of the quartz grains are monocrystaline of the normal type. Polycrystalline grains are rare. The bioclast percentage is around 10 % and consists of mainly from molluscan fragments (Fig. 3.15a, b). These fragments were originally aragonite and HMC. Later it is enveloped with micrite and then the aragonite dissolved and the shells calicitized by sparry calcite (LMC) (Fig. 3.15c). Other detrital minerals are represented by k-feldspar. The cement is represented by patches from calcite and replaced by equant rhombic siderite or ferron dolomite (Fig. 3.15d). The original rock could be carbonate rock and replaced by siderite. Some siderite rhomb altered to limonite and this can be seen clear in the reflected light. In addition, feldspar minerals are altered to sericite.

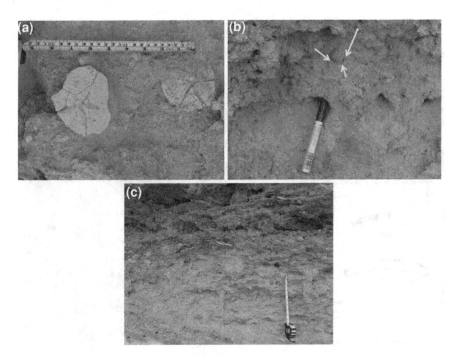

Fig. 3.14 Photographs of common lithofacies in bioturbated–fossiliferous sandstone facies association (FA5). **a** Close-up view of *Scutella* in fossiliferous limestone. **b** Close up view of bivalve shell (look the *arrows*). **c** Bioturbated fossiliferous sandstone

Fossiliferous Ferruginous Large Scale Cross-Bedded Sandstone with Hard Crust (F13)

Description

This facies ranges in thickness between (2.5 m, unit II, section 1 and 3 m, bed 2-2-7, unit II, section 2) and consists of coarse to medium grained sand with two hard crusts (Fig. 3.16a). The two hard crusts consist of laminated fine grained sand. The crusts are highly cemented by iron and manganese. In-between these two hard crusts the bed is formed of yellowish color poorly sorted sand.

A characteristic feature of this facies is the presence of tangential cross beds that are ebb-current directed, 330–310° (Fig. 3.16b, c). It shows compound cross-stratification on a large scale (2 m thick). These cross sets are characterized by superimposed smaller scale (0.3–0.15 m thick) minor cross lamination parallel to the large sets. The upper part of this facies contains shells from pelecypods. This facies is of limited extension and just restricted to two sections 1 and 2 and it is overlain by tabular and trough cross-stratified sandstone (F4) and is underlain by channel in fill deposits (F1).

Petrography is mainly from pebbly calcareous quartz arenites which it is consists of poorly sorted sand. The dominant of quartz is moncrystaline and few grains are composite or polycrystalline (Fig. 3.17a, b). The composite nature of the grain

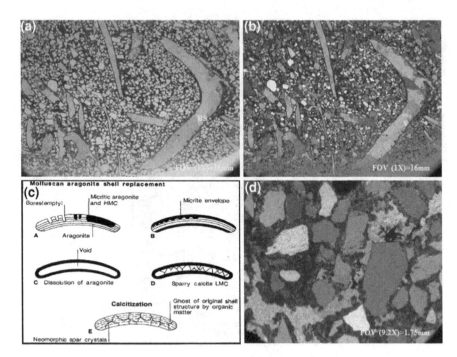

Fig. 3.15 Photographs of common lithofacies in bioturbated–fossiliferous sandstone facies association (FA5). Sidertize bioclastic bearing quartz Arenites. Note coated bivalve shells (BS) and ostracodis represent by disarticulated valve (*blue arrow*). All the primary aragonitic mollusk shells are dissolved. **c** Molluscan aragonite shell replacement (Tucker et al. 1990). **d** The cement around the quartz grains is composed wholly of mintuezoned rhomoherdra of dolomite. Some of the quartz grains are partially eaten by dolomite probably during dolomitization

is clear only in the view taken with polars crossed. Note that the boundaries between the crystals are straight. It is known that the composite quartz from igneous sources usually has straighter crystal boundaries. Regarding to packing it is open packing and poorly sorted. The glauconite grains are present with percent less than %1 and these grains have a characteristic yellowish color and some of them featureless (Fig. 3.17c, d). Feldspars are relatively more common than in the accompanying the quartz arenites. Some of the feldspar grains may contain inclusions. Evidence of alteration is occasionally observed which it is altered to sericite. Cement from calcite rhomb with high interference color.

Glauconitic Trough Cross-Bedded Sandstone (F14)

Description

It found in (1 m, bed 21-9-1, unit IX, section 21). It started with 25 cm lag deposits with mudclasts in calcareous sandstone matrix. The whole facies is yellowish red color and coarse grain glauconitic sandstone. The sandstone is characterized by cross stratification. The cross stratification is represented by troughs with dominate direction (350°) and the cross lamina 275° (Fig. 3.18). In addition, this facies is characterized by lenses of sandstone considered as markers for this

Fig. 3.16 Photographs of common lithofacies in bioturbated–fossiliferous sandstone facies association (FA5). **a** Intensively bioturbated glauconitic sandstone with two hard crusts (*arrows*). **b, c** Close-up view of tangential cross-beds that are ebb-current directed, 330–310°and it shows compound cross-stratification on a large scale

facies. This facies is overlain by bioturbated heterolithic sandstone (F19) and is underlain by tabular and trough cross stratified sandstones (F3). This facies can be presented as intensively bioturbated glauconitic sandstone and found in (section 4, 1.8 m, unit I, beds no. 4-1-8). It consists of slope forming yellow fine to medium grained quartzose sand rich in green glauconitic pellets, moderately to well sorted with thin streaks of shale. This facies is distinguished by wavy bedding (lenticular shape) with mud drapes over it. This facies is overlain by sand-shale intercalation (F17) and is underlain by bioturbated heterolithic sandstone (F19).

Interpretation (F10–F14)
Figure 3.19 is a modified from Harris et al. (1997) (after Lindholm 1987) ternary diagram illustrating the proposed energy regimes and interpreted depositional environments for the lithofacies types identified in our study. Quartz sand (S) in Fig. 3.19 represents high-energy conditions in a nearshore environment supported by the presence of glauconite and comminuted fossil debris including well-rounded pelecypods, echinoderms, foraminifers, and gastropods. Calcareous sand (cS) represents moderate- to high-energy transitional environments from nearshore to inner shelf.

The lack of terrigenous mud and presence of broken and rounded echinoderms and mollusks is supportive of these environments Harris et al. (1997).

The presence of glauconite associated with the faunal assemblage of pelecypods, and other skeletal fragments indicates a marine depositional environment

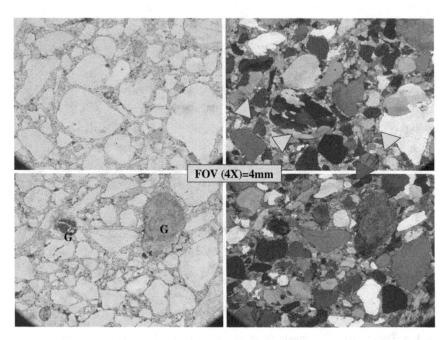

Fig. 3.17 Thin section photomicrographs of common lithofacies in bioturbated–fossiliferous sandstone facies association (FA5), pebbly calcareous quartz arenites. **a, b** Poorly sorted quartz arenites. Note the polycrystalline quartz grain (*yellow arrows*) and feldspar particle (albite) showing multiple twinning (*red arrow*). The much finer sediment surrounds the composite quartz grains contain moncrystaline quartz, PPL, XPL. **c, d** The rounded glauconite grain shown near from the center and seems reworked

of normal salinity that probably accumulated on the inner to middle shelf (Milliman 1972).

The trace fossil *Thalassinoides* commonly represents a dwelling or feeding structure produced by a decapod crustacean, the Thalassinid shrimp (e.g., Pemberton et al. 2001). This trace is typically related to lower shoreface environments. On the basis of the presence of *Thalassinoides, Glossifungites* Ichnofacies assemblages, *Thalassinoides* is common in lower marine shoreface to offshore environments, but it also is known to occur in brackish water environments associated with low ichnogneric diversity (Pemberton et al. 2001). *Thalassinoides* is regarded as a dwelling/feeding burrow of a deposit-feeding Thalassinid shrimp (Myrow 1995) and is sometime found to have a marginal marine affinity (Gingras et al. 2002). Extensive burrows suggest deposition under marginal marine conditions (e.g., de Raaf and Boersma 1971; Nio and Yang 1991a).

Ophiomorpha is common in marine sandy substrates, and elaborate burrow systems often are prolific in shoreface environments (Frey et al. 1978). Occurrence in brackish water settings includes estuaries and tidal shoals. These dwelling burrows were created by decapods crustaceous, most commonly callianassid shimp (Frey et al. 1978; Pemberton et al. 2001 cited in Belt et al. 2005).

Fig. 3.18 Photographs of common lithofacies in bioturbated-fossiliferous sandstone facies association (FA5). Close-up view of glauconitic trough cross -bedded sandstone. The arrows indicate different size for the cross-bedded forsets

Pectinids occur frequently and are widely distributed in shallow marine environments. Moreover, distinct morphological groups developed different life strategies (Waller 1991).

Glauconite is a seafloor diagenetic product formed primarily in mid- to outer-shelf settings, specifically during intervals of slow sedimentation rate and mildly reducing conditions. Fecal pellets of a variety of organisms provide an excellent site for neoformation of glauconite because the localized reducing environment that exists in the presence of residual fecal organic matter (see, Scholle and Ulmer-Scholle 2003). The occurrence of abundant glaucony at particular horizons in stratigraphic sequences is usually interpreted to be the result of slow accumulation due to sediment starvation, typically in water depths ranging from 50 to 500 m (Amorosi 1997; Odin and Fullagar 1988). Thus, in sequence stratigraphic analysis glaucony-rich beds can serve as convenient indicators of either transgressive or maximum flooding surfaces (e.g., Baum and Vail 1988; Galloway 1989; Loutit et al. 1988). In detail, however, rather diverse patterns of glaucony abundance and maturity through shallow-water shelf sequences have been documented recently (e.g., Huggett and Gale 1997; Amorosi 1997; Amorosi and Centineo 1997; Kitamura 1998; Stonecipher 1999). Moreover, when deeper-water settings have been considered, the sequence stratigraphic significance of glaucony accumulation appears even more complex, with different levels through a stratigraphic section inferred to represent all conceivable states of relative sea-level change (Mccracken et al. 1996; Miller et al. 1998) (cited in Hesselbo and Huggett 2001).

Glaucony is commonly associated with condensed sections, unconformities and transgressive deposits (Amorosi 1995; Loutit et al. 1988; Shanmugam 1988), and

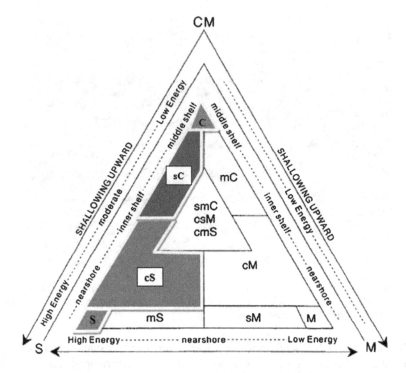

Fig. 3.19 Modified ternary diagram of Harris et al.1997 (after Lindholm 1987) showing interpreted environments of deposition and energy levels for carbonate terrigenous lithofacies examined in this study. *CM* carbonate mud, *S* terrigenous sand, *M* terrigenous mud. Fields within the *triangle* include: *S* terrigenous sand; *cS* calcareous sand; *SC* sandy carbonate; *c* carbonate; *mC* muddy carbonate; *smC* sandy, muddy carbonate: *CSM* calcareous, sandy mud; *cmS* calcareous, muddy sand; *cM* calcareous mud; *mS* muddy sand; *sM* sandy mud; *M* terrigenous mud

as such may contribute to an increased understanding of sea-level cyclicity (Kelly and Webb 1999).

Formation of Glauconite and the Glaucony Facies (Hesselbo and Huggett 2001). The "glaucony facies" comprises ferric-iron-rich, glauconitic minerals with a 10–14 A° basal lattice spacing, typically occurring as green to greenish-brown pellets (Odin and Matter 1981). "Glauconite" is the green, potassium- rich end-member mineral of this grouping, and has a 10 A° basal lattice spacing. Glaucony forms in areas of slow sedimentation where there is a suitable substrate, a semiconfined, suboxic environment, and an abundant supply of iron. The principal substrates for glaucony are fecal pellets, tests, and phyllosilicate grains. A well-known classification system for glaucony has been devised by Odin and Matter (1981): nascent, slightly evolved, evolved, and highly evolved. Glaucony formation commences just beneath the sediment–water interface, with the formation of iron-rich smectitic clay; as glauconitization proceeds, evolved grains develop surface cracks, which may then themselves become infilled with pale green glaucony. If the infilling glaucony also undergoes maturation the pellets become uniformly highly evolved grains (Odin and Dodson 1982). With increasing maturity the potassium content increases, and the basal lattice spacing decreases. Evolved glauconite may be formed in 105 years, if the granules are not buried (Giresse et al. 1980). Chemical evolution (uptake of Fe and K) stops either after long exposure at the sediment–water interface or after burial to several decimeters (Odin and Matter 1981; Odin 1988). Hence glaucony is commonly associated with marine transgressions, where rapid deepening starves the shelf of sediment, and maturity of glaucony may be an indicator of the intensity of a hiatus. Potassium-rich evolved glaucony continues to mature during burial diagenesis with substitution of aluminum for iron (Ireland et al. 1983).

The presence of glaucony by itself is not diagnostic of a specific systems tract of a depositional sequence. A dependable sequence stratigraphic interpretation of glaucony-bearing units requires additional information on glaucony, including: (1) spatial distribution, (2) genetic attributes (differentiation of autochthonous from allochthonous, and intrasequential from extrasequential glaucony), and (3) maturity (distinction between nascent, slightly evolved, evolved, and highly evolved glaucony).

Autochthonous glaucony is common at various stratigraphic levels in the transgressive systems tract (TST) and the lower highstand systems tract (HST), showing an upward increase (TST) and then decrease (HST) in abundance and maturity. The condensed section can be distinguished from the overlying and underlying deposits by more concentration and maturity of glaucony. Allochthonous intrasequential (parautochthonous) glaucony can be found in the entire TST, HST, and lowstand systems tract (LST), generally showing less concentration and maturity than its autochthonous counterpart. Allochthonous extrasequential (detrial) glaucony is found mainly in the LST, its concentration and composition depending on the characteristics of the source horizon. The connection between autochonous and allochthonous (intrasequential and extrasequential) glaucony commonly exists in the LST and in the lower TST (Amorosi 1995).

Bioturbated: Fossiliferous Carbonate Facies Association (FA6)
Fossiliferous Limestone (F15)

Description

This facies is limited to the northern and western part of the area in sections (5, 5′, 6, 7, 9, 10, 21 and 22). It accumulated maximum thicknesses in section 10 about 9 m and consists of white color sandy fossiliferous limestone (Fig. 3.20a–d). It can be divided into six distinct micro-facies (F): The most common miro-facies are mud dominated packstone, quartz skeletal mud dominated dolo-packstone, siliciclastic grain-dominated wackestones to siliciclastic grain-dominated packstones, siliciclastic grain-dominated packstones, quartz peliodal dolo-grainstone and quartz rich caliches or peletal fabric with caliches. These carbonate sediments that represented by fossiliferous limestone are constructed exclusively of skeletal remains, notably bryozoans, bivalves, red algae, benthic foraminifers and barnacles. In addition, non-skeletal carbonate grains are represented by peloids and few aggregates. For more details on fossiliferous limestone subdivision see Chap. 4.

Fig. 3.20 Photographs of common lithofacies in bioturbated-fossiliferous carbonate facies association (FA6). **a, b** Fossiliferous limestone. **c** Abundant rip-up clasts in the base of this fossiliferous limestone. **d** Close-up view of echinoid shell in fossiliferous limestone

Interpretation (F15)
Fossiliferous limestone is interpreted to be the most seaward facies in the Moghra formation. The diverse fauna of all facies suggests deposition in clear, well-oxygenated, open-marine water of normal salinity on the inner to middle shelf (∼30 m depth) with periods of marginal marine, beach, and deltaic influence. All the above criteria support open-marine water of normal salinity on the inner to middle shelf for this facies association (Harris et al. 1997). For more details on fossiliferous limestone subdivision interpretation see Chap. 4.

3.2.3 Tide-Dominated Delta

Coarsening Upward Facies Association (FA7)
Coarse Grained Lensoidal Heterolithic (F16)
Description
This facies is represented by 4 m (bed 21-9-4, unit IX, section 21) of thinly laminated siltstone and mudstone intercalated with fine grained sandstone (Figure 3.21a–c). This bed grades up into fine grained sand (Fig. 3.21d). The uppermost part of this bed (1 m from the top) is represented by 50 cm different sized lenses of coarse-grained quartz. In other units (unit XIII, bed 21-15-6, section 21)

Fig. 3.21 Photographs of common lithofacies in coarsening upward Facies association (FA7). **a, b** Close up view of thinly laminated siltstone and mudstone intercalated with fine grained sandstone. **c, d** Coarsing upward heterolithic bed

this intercalation from silt and sand is represented only by 3.5 m mudstone with two iron crust and ended also by lens from coarse grained sand (50 cm). This facies is overlain erosively by facies (F1) of axial transgressive, tide-influenced fluvial channel (FA1) and underlain by glauconitic trough cross-bedded sandstone (F14).

Thin Laminated Sand: Shale Intercalation (F17)

Description

This facies is represented by 2 m (bed 21-1-1, section 21) thin laminated, fissile grey shale with highly ferruginated very fine grain quartzose sand (Fig. 22a). The sandy laminae become dominant upwards. The shale is more clayey in the base and coarsens to become more silty in the upper part (Fig. 3.22b), no burrows, no plant remains, abundant shale streaks are recorded. This facies is overlain by homogenized bioturbated sandstone (F23) which coarsens upwards. The base of this unit is unexposed in this section but other sections (7 and 8) it is underlain by sandstones with moderately to intensively bioturbated branched network *Ophiomorpha* (F10).

Homogenized Bioturbated Sand (F18)

Description

This facies is represented by 0.6 and 2,4 m (bed 7-3-6, section 7 and bed 21-1-5, section 21) quartzose sand, pebbly in places, poorly sorted, homogenized

Fig. 3.22 Photographs of common lithofacies in coarsening upward Facies association (FA7). **a** Thin laminated, fissile *grey* shale with very thin ferrugineous very fine grain quartzose sand. **b** The shale is more clayey in the base and coarsens to become more silty in the upper part. **c** Homogenized bioturbated sand. **d** Bioturbated sand

(Fig. 3.22c) and bioturbated (Fig. 3.22d). This facies is shaley at the base and becomes sandier in the upper part. This facies can be represented also in (unit 1, bed 1-1-3-2, section 1) by 140 cm massive homogenized bioturbated sandstone with reworked fragment of silicified wood. The grain size is coarse and medium size. This bed is slightly cemented with argillaceous cement. This facies is overlain by inclined heterolithic stratification (F6, see section 7) and underlain by mudstone beds (F8).

Bioturbated Heterolithic Sandstone (Interbedded Mudstone and Sandstone) (F19)

Description

This facies is represented by 2.4 m to 3 m (beds 21-1-10 and 21-9-2, section 21) shale - sand intercalation (Fig. 3.23a). The shale is dominant at the base with increasing sand upwards (Fig. 3.23b, c). This facies is moderately bioturbated. In addition, it characterized by gypsum streaks parallel to the bedding plan and other cross cutting the beds. This facies is changed vertically and represented by 3 m (21-9-2, section 21) ended by coarse grain sand which bioturbation not marked this bed. The interbedded mudstone and sandstone displays a cyclical stacking pattern of coarsening and thickening upward. The top part is characterized by

Fig. 3.23 Photographs of common lithofacies in coarsening upward Facies association (FA7). **a–c** Moderately bioturbated interbedded mudstone and sandstone displays a cyclical stacking pattern of coarsening and thickening upward. The head of *triangle* indicates that the shale is dominant at the base with increasing sand upwards. Photographs of common lithofacies in fining upward Facies association (FA8). **d** Tabular and trough cross-stratified, mainly fine grained quartz sand

ferruginous erosional surfaces with mudclasts and marked with lag gravel. This facies is overlain by cross stratified sandstone with mudstone drapes (F20) and is underlain by glauconitic trough cross-bedded sandstone (F17).

Interpretation (F16–F19)
Based on the significant sedimentary thickness (24 m) (max thick in section 20) and the coarsening upward trend, a generalized delta-front depositional setting has been interpreted for this facies association. The studied sedimentary features of this facies association 4 are in agreement with this generalized interpretation (Olivero et al. 2008). In particular, the coarsening and thickening upward arrangement of sedimentary facies are consistent with a deltaic depositional setting. Within the context of this depositional environment, the coarsening upward packages of tidal flat facies association FA4 are interpreted to reflect prodelta and delta front settings (Dalrymple and Choi 2007; Olivero et al. 2008). The delta-front and prodelta areas contain a seaward decreasing amount of sand and generate an upward-coarsening succession during progradation. The delta-front deposits consist of interbedded sand and mud, in which the mud may be structureless and unbioturbated because it was deposited rapidly by fluid muds (fluid-mud deposits i.e., structureless mud layers more than 0.5–1 cm thick that were deposited in a single slack-water period,

Dalrymple and Choi 2007). Slow, passive settling of fine grained sediment from suspension is also possible (Harris et al. 2004).

The deposits of prodelta mudstone record sedimentation from suspension in low energy settings. The sandier, upper part of each package suggests distributary mouth bars prograding onto distal siltstones and mudstones. Dominance of massive and parallel-laminated sandstone probably reflects inertia-dominated deposition of fine sand grains introduced into the basin during times of high fluvial discharge (Bates 1953; Bhattacharya and Walker 1992). Tidal rythmites may be present (Jaeger and Nittrouer 1995), but long successions (more than a few days) are unlikely because of disturbance by waves (Dalrymple and Choi 2007).

The general increase in sand content up-section reflects progradation of the delta front (cf. Bhattacharya 2006; Elliott 1986; Olivero et al. 2008). The lack of fluvial delta plain deposits between successive packages probably reflects the distal character of the subaqueous site or erosion by subsequent channels.

Fining Upward Facies Association (FA8)
Cross-Stratified Sandstone with Mudstone Drapes (F20)
Description
This facies consists of 6 m thick (unit 1, bed 8-1-1, section 8), thin laminated, flazer laminated, tabular and trough cross-stratified (Fig. 3.23d), mainly fine grained quartz sand, light colour with highly ferruignated thin laminae at the base. The fine-grained sand is highly bioturbated by *Ophiomorpha* burrows, mainly vertical burrows. The bioturbation makes the bed mottled brown. This facies sometimes is represented by 1.5 m thick, yellowish color sand, very fine to fine grain quartzose, micaceous and green colored mud flakes. The top most part is highly cemented, coarse grained, pebbly with abundant mudclasts, the upper surface is sharp undulate with shark teeth and shell fragments. The top most part is cross bedded with foresets towards 185$\mathring{}$. This facies is overlain by thin laminated sand-shale intercalation (F17) and base of this faces unexposed.

Interpretation
The sigmoidal reactivation surfaces and the cyclic thickening and thinning of cross strata, together with the abundant mudstone drapes, suggest deposition from tidal currents (e.g., Kreisa and Moila 1986; Nio and Yang 1991b; Shanley et al. 1992). The significant height and length of the sandstone bodies together with the bidirectional paleocurrent directions and the abundant tidal indicators suggest deposition as subtidal bars on the delta front reaches of the system (see also Pontén and Plink-Björklund 2009). The bars also have more pronounced lateral accretion surfaces and predominantly landward paleocurrent directions. Tidal bars have been described from deltaic (Maguregui and Tyler 1991; Mccrimmon and Arnott 2002; Willis 2005; Willis et al. 1999; Willis and Gabel 2003) and shelf (Allen 1980; Dalrymple 1992) settings. The association with the supratidal-flat mudstones in the landward end, together with the bioturbated and wave-influenced basinward parts of the tidal bars suggests that the bars extended from a protected coastal nearshore setting (see Dalrymple et al. 1990) to a more open marine setting, where wave reworking occurred.

A Comparison Between Tidal Bars from Deltaic Versus Estuarine Settings
Estuarine tidal bars are characterized by: (1) preserved bar topsets; (2) dominantly
landward palaeocurrent directions; (3) well to very-well-sorted, very fine- or fine-
grained sandstones; and (4) overall retrogressive stacking evidenced by fluvial-
tidal bars overlain by marine estuarine-mouth bars and then by overlying marine
mudstones. These bars, were deposited in the estuary, as elongated sand bodies.
Tidal currents dominated the depositional regime, seen from the abundant tidal
facies, and high energy is suggested by low abundance of mudstone drapes and
mica drapes. The deltaic tidal bars are characterized by: (1) eroded bar topsets; (2)
dominantly basinward (ebb) palaeocurrent directions; (3) poorly sorted sandstone
with grain-size ranging from very fine sands to granules; and (4) overall progradational stacking patterns. These tidal bars were situated seaward of the delta
plain. Strong fluvial influence is reflected by coarse-grained, poorly sorted sandstone, mudstone and pebble clasts and the predominately basinward palaeocurrent
directions (Pontén and Plink-Björklund 2009). Of great importance is that deltas
are muddy at their seawards end with a characteristic prodelta facies at their base,
whereas estuaries tend to have no mud at their seaward end, but rather end seawards in an open marine sandy facies.

3.2.4 Depositional Environment of Moghra Formation

The present observations on the paleoenvironmental settings of the Early Miocene
Moghra Formation have resulted in three inter-fingering environmental interpretations: it is likely that the Moghra Formation was deposited in three environments: (1) Tide-dominated Estuary (Facies Association 1), (2) Open Shelf (Facies
Association 2), and (3) Tide-dominated Delta (Facies Association 3). Facies
Association 2 was broadly contemporaneous with and formed basinward of both
Facies Associations 1 and 3. Facies Associations 1 and 3 would have occupied the
same position (i.e. the fluvial to marine transition) at different times. The lower
part of (2) in any outcropping facies succession would genetically relate to (1),
whereas the upper part of (2) would be broadly co-eval with (3).

The paleoenvironmental conclusions are based partly on paleoecological
interpretations. Sedimentary facies and architecture favor depositional settings
related to tidal channels and tidal bars of various scales and paleogeographic
positions within an estuary and its genetically related, underlying delta.

3.2.5 Tide-Dominated Estuaries, Open Shelf and Deltas of the Moghra Formation

Tide-Dominated Estuary

The estuary environment of Moghra area contains two facies associations: (a)
axial, high-energy tidal channels with infilling bars/dunes that proximally become

tidal-fluvial channels and (b) marginal, low-energy intertidal to supratidal flats (as is common most modern estuaries), from near the mouth of the estuarine funnel to the landward limit tidal influence.

The Axial estuary zone is subdivided into: Inner-most estuary, tide-influenced fluvial channels; Mid-estuary, upper-flow regime sand-flats; and Outer-estuary tidal bars and channels.

Innermost estuary, tide-influenced fluvial channels to Mid-Estuary flat-laminated sandstone

In this study, fluvial-tidal deposits are defined as deposits of a predominantly fluvial source with a tidal signature and were deposited under mixed fluvial and tidal conditions. The fluvial-tidal zone starts at the landward limit of the estuary and ends at a certain point in the estuary; obviously its limits constantly change through time. Developed diagnostic criteria for the fluvial-tidal zone (compared to mid and outer estuary areas) are a greater number of erosional surfaces, abundance of lag deposits, especially tree trunk logs and vertebrates and tabular and trough cross stratified sandstones with less marine bioturbation. Sandy channel deposits formed under mixed brackish and fresh water conditions, but with mud drapes formed during tidally induced flow reversals.

Outer Estuarine Bars and Channels

The sand in the outer estuarine bars is coarser grained than in mid-estuary and generally allowed the development of widespread, subaqueous dune fields. Paleocurrents are predominantly landward directed, because of the flood dominance of this area, but with some areas of ebb dominance. Trace fossils are rare in the sandy sediments because of constant sediment movement. However, in more distal settings, such as in the transition to the shelf where sediment movement may be more intermittent, bioturbation is more pervasive (cf. Harris et al. 1992). The detailed characteristics of the deposits are as follows: We distinguished two tidal facies, (1) Tabular and trough cross stratified sandstones and lensoidal calcareous cross-stratified sandstone with *Ophiomorpha* burrows, and (2) marginal, low-energy intertidal to supratidal flats (as is common most modern estuaries) with inclined heterolithic stratification, flat-lying heterolithic beds, mudstone beds and mangrove-bearing cross bedded sandstone. Intertidal deposits show strong bioturbation.

So, we can support our present work for these criteria that dominated in estuary dominated as in Heap et al. (2004). As a result, elongate tidal channels characterised by strong tidal currents develop throughout the system. In macrotidal systems, the funnel-shape of the estuary is generally preserved into the delta stage (Dalrymple et al. 1992).

Open Shelf (Tide-Dominated Shelf)

These deposits consist of bioturbated-fossiliferous carbonate that can be subdivided into fossiliferous limestone, moderately bioturbated fossiliferous limestone and sandy fossiliferous limestone. Highly fossiliferous limestone is rich in scutella, shell fragments, molds of pelecypod and gastropod. The base of this tract is lined by coarse grained, pebbles and mudclasts. Bioturbation is moderate to high and includes complex sub horizontal to inclined, branching burrow networks, including *Ophiomorpha*. *Ophiomorpha* is characterized by branching, three-dimensional burrow systems that comprise shafts and tunnels containing well developed linings of agglutinated sediment. These extensive burrows are commonly lined with iron oxide cement. Burrow linings are smooth on the interior and they are characteristically pelleted on the exterior. Burrow systems range from simple to complex, with irregularly spaced, Y-shaped branches. A characteristic feature of these deposits are the presence of tangential cross beds that are ebb-current directed, 330–310° and the compound cross-stratification on a large scale (2 m thick). These cross sets are characterized by superimposed smaller scale minor cross lamination parallel to the large sets. The shelf area seaward of the outer-estuarine tidal bars has a different character than the delta-front and prodeltaic area that lies seaward of deltaic systems that described below.

Tide-Dominated Delta
These deposits (described mostly from hand-dug trenches) are initially a basal, bioturbated marine shale and that passes upwards to heterolithic deltaic strata to include unbioturbated, unlaminated mud layers of fluid-mud origin. The transition shows a very irregular upward grain-size coarsening, bed thickening, and increasing sandstone content. Sandstones in the heterolithic strata are very fine to medium-grained and some sandstone beds are cross stratified. The gradual transition from the basal shale to the heterolithic strata suggests progradation of delta-front environments across prodelta environment.

Characteristic Signals of Tidal Processes and Occurrence in Moghra Succession
Several authors have discussed basic principles of tides (Friedman and Sanders 1978; Strahler and Strahler 1974). Most coasts and continental shelves are subjected to semi-diurnal tides (i.e. rises and falls in sea level twice daily). Tidal currents are very effective in transporting and depositing sand in estuaries where velocities commonly reach 150 cm/s (Reading and Collinson 1996). Sedimentary features indicative of tidal processes in shallow-water environments (Allen 1982; Archer 1998; Banerjee 1989; Dalrymple 1992; Klein 1970; Nio and Yang 1991a; Reineck and Wunderlich 1968; Shanmugan et al. 1998, 2000; Terwindt 1981; Visser 1980) are: (1) Heterolithic facies, (2) rhythmic alternation of sandstone-shale couplets (tidal rhythmites), (3) thick (spring)-thin (neap) bundles of foreset strata in dunes, (4) double mudstone layers representing the two slack-water intervals within a single tidal cycle, (5) cross-beds with mud-draped foresets, (6) bidirectional (herringbone) cross-bedding, (7) sigmoidal cross-bedding (i.e. full-vortex structures) with mud drapes and tangential basal contacts, (8) reactivation surfaces (not unique for tides, can be found in fluvial), (9) elongate mudstone clasts, (10) flaser bedding, (11) wavy bedding, and (12) lenticular bedding (Shanmugam 2003). Furthermore, from the ichnology point of view there are two recurring ichnofossil assemblages that show that marine and brackish waters, not freshwater, were present during facies development. These are: (1) a brackish-water to marine, *Thalassinoides, Ophiomorpha* ichnofabric

(FA3); and (2) a brackish water, *Thalassinoides*-generated ichnofabric that is interpreted to have descended into consolidated substrates, and thus represents the *Glossifungites* Ichnofacies (FA3). Although absolute salinity is difficult to assess from trace-fossil data, abundant ichnological and neoichnological evidence show that tubular, smooth-walled, regularly branching *Thalassinoides* are a marginal-marine to marine phenomena (Dworschak 1983; Frey and Howard 1975; Frey et al. 1978; Gingras et al. 2000; Griffis and Chavez 1988; Rice and Chapman 1971; Shinn 1968). Modern studies show that *in firmgrounds,* these trace fossils are made by a thalassinid shrimp (Gingras et al. 2000, 2001), which indicates that sedimentation occurred in mesohaline to marine waters (Gingras et al. 2002).

From these above evidence it is strongly suggested that the Moghra succession is dominated by tidal processes and that individual half-cycles represent individual tide dominated estuaries.

Tide-dominated estuarine successions (Fig. 3.24) are distinguished from wave-dominated tidal estuary succession by the following criteria: (1) the development of tidal sand bars and upper-flow-regime sand flats instead of barrier and shoreface deposits, (2) the development of tidal meanders and salt marsh instead of a central basin (lagoon), (3) the absence of a bay-head delta, and (4) an erosive ravinement surface formed by tidal currents discontinuously and irregularly penetrates back into the estuary instead of there being only a wave-ravinement surface created by waves at the mouth of the estuary Dalrymple et al. (1992). Although the tide-dominated model seems reasonable and is widely accepted today, it validity still requires verification with more ancient examples (Kitazawa 2007).

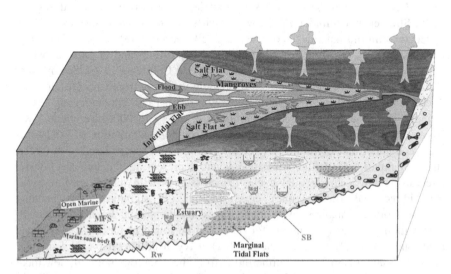

Fig. 3.24 Schematic section of a depostional sequence formed in a tide-dominated estuary incised-valley system during transgression-regression cycle, *SB* sequence boundary; *Rw* tidal ravinement surface; *MFS* maximum flooding surface

3.2.6 The Broader Setting of the Moghra Estuaries

Dalrymple (2006), Dalrymple and Choi (2007), and Løseth et al. (2009) expanded
the earlier Dalrymple et al. (1992) definition of estuaries to include transgressive
coastal areas receiving fluvial and marine sediment input also in areas without
incised valleys (i.e. abandoned delta plains). In the case of the Moghra succession
the estuarine units usually always lie within significant erosional incisions (up to
20 m deep), but this does not necessarily mean that each of the estuarine units
represents an individual incised valley. We know that the distributaries channels of
many tide-dominated deltas can cut down to 10 s of meters by normal tidal-current
scour without there being a base-level fall involved or without the erosional surfaces being of wide regional extent (Willis and Gabel 2001, 2003). These deep
erosional scours therefore may or may not be sequence boundaries, and would then
have been back-filled by individual Moghra estuaries. On the larger scale of the
entire Moghra succession the question still remains whether or not the entire
estuarine complex fills a larger master valley or not. Thus there are two alternatives, one involving a very large, master incised valley, and the other a non-
valleyed scenario for the Moghra estuarine-delta system (Figs. 3.24 and 3.25).

**Alternative Valleyed and Non-valleyed Interpretations of Moghra
Estuarine Complex**
Early definitions of estuaries focused on oceanographic parameters, such as
salinity (e.g., Pritchard 1967) and tides (e.g., Fairbridge 1980). Later, a widely
used facies model, focusing on the co-occurrence of landward sediment transport
and marine transgression, was proposed by Dalrymple et al. (1992). This model

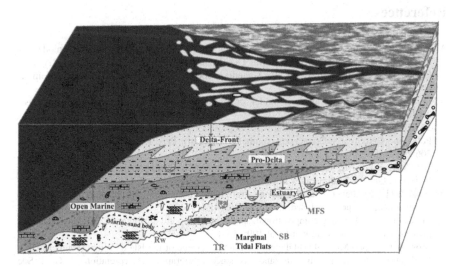

Fig. 3.25 Schematic section of a depositional sequence formed in a tide-dominated delta incised-
valley system during transgression-regression cycle, *SB* sequence boundary; *Rw* tidal ravinement
surface; *MFS* maximum flooding surface

also linked the definition of an estuary to the erosive container (the valley) within which the estuarine sediments were deposited. In this model, estuaries occupy palaeo-valleys formed by fluvial erosion during fall and lowstand of relative sea level (incised valley). This relationship was further elaborated upon in models describing incised-valley depositional systems (e.g., Dalrymple et al. 1994; Zaitlin et al. 1994).

The estuarine definition of Dalrymple et al. (1992) appears now to have been too restrictive. Although estuaries are most common within incised valleys, several studies have suggested that they can have other confinements or no 'confinement' at all in the sense of a narrow container (e.g., Carr et al. 2003; Jackson et al. 2005; Porebski 2000). This led Dalrymple (2006) to a new, modified definition of an estuary, where the original reference to the necessity of incised valley is removed. He then stated that "an estuary is a transgressive coastal environment at the mouth of a river, that receives sediment from both fluvial and marine sources, and which contains facies influenced by tide, wave and fluvial processes. The estuary is considered to extend from the landward limit of tidal facies at its head to the seaward limit of coastal facies at its mouth" (Dalrymple 2006).

In the present study, we document a depositional system that has all the characteristics, in terms of depositional facies, of a well-developed, repeatedly-stacked, estuary-delta system. However, the Moghra system possibly lacks evidence of an obvious sub-aerial unconformity at the erosive base of the entire Moghra complex. For this reason the non-valleyed alternative is a real possibility and would be consistent with Dalrymple's (2006) recent modification of the definition of an estuary (Løseth et al. 2009).

References

Allen G (1991) Sedimentary processes and facies in the Gironde estuary: a recent model for macrotidal estuarine systems. In: Smith DG, Reinson GE, Zaitlin BA, Rahmani RA (eds) Clastic tidal sedimentology, vol 16. Canadian Society of Petroleum Geologists, Memoir, pp 29–39

Allen JRL (1980) Sand waves: a model of origin and internal structure. Sediment Geol 26:281–328

Allen JRL (1982) Mud drapes in sand-wave deposits: a physical model with application to the Folkstone Beds (early Cretaceous, southeast England). Proc Royal Soc Lond, v. Series A 306:291–345

Allison MA, Khan SR, Goodbred SL, Kuehl SA (2003) Stratigraphic evolution of the late Holocene Ganges-Brahmaputra lower delta plain. Sediment Geol 155:317–342

Amorosi A (1995) Glaucony and sequence stratigraphy: a conceptual framework of distribution in siliciclastic sequences. J Sediment Res B65(4):419–425

Amorosi A (1997) Detecting compositional, spatial and temporal attributes of glaucony: a tool for provenance research: Sediment Geol 109:135–153

Amorosi A, Centineo MC (1997) Glaucony from the Eocene of the Isle of Wight (southern UK): implications for basin analysis and sequence-stratigraphic interpretation. J Geol Soc 154(5):887–896

Archer AW (1998) Hierarchy of controls on cyclic rhythmite deposition: carboniferous basins of eastern and mid-continental USA. In: Alexander CR, Davis RA, Henry VJ (eds) Tidalites:

processes and products, vol 61. SEPM Society for Sedimentary Geology Special Publication, pp 59–68

Banerjee I (1989) Tidal sand sheet origin of the transgressive basal Colorado Sandstone (Albian): a subsurface study of the Cessford field, southern Alberta. Bull Can Pet Geol 37:1–17

Bates CD (1953) Rational theory of delta formation. Am Assoc Pet Geol Bull 37:2119–2162

Baum GR, Vail PR (1988) Sequence stratigraphic concepts applied to Paleogene outcrops, Gulf and Atlantic basins. In: Wilgus CK, Hastings BS, Kendall CGStC, Posamentier HW, Ross CA, Van Wagoner JC (eds) Sea-level changes: an integrated approach. SEPM, Special Publication 42:309–327

Belt ES, Tibert NE, Curran HA, Diemer JA, Hartman JH, Kroeger TJ, Harwood DM (2005) Evidence for marine influence on a low-gradient coastal plain: ichnology and invertebrate paleontology of the lower Tongue River Member (Fort Union Formation, Middle Paleocene), Western Wiliston Basin, U.S.A. Rocky Mount Geol 40(1):1–24

Bhattacharya J, Walker RG (1992) Deltas. In: Walker, RG, James, NP (eds) Facies models: response to sea level change. Geological Association of Canada, St. John's, pp 157–177

Bhattacharya JP (2006) Deltas. In: Posamentier HW, Walker RG (eds) Facies models revisited, vol 84. SEPM (Society for Sedimentary Geology) Special Publication 84:237–292

Boersma J, Terwindt J (1981) Neap-spring tide sequences of intertidal shoal deposits in a mesotidal estuary. Sedimentology 28:151–170

Bown T, Kraus M (1988) Geology and paleoenvironment of the Oligocene Jebel Qatrani Formation and adjacent rocks: Fayum Depression Egypt. US Geol Surv Prof Pap 1452:60

Carr I, Gawthorpe R, Jackson C, Sharp I, Sadek A (2003) Sedimentology and sequence stratigraphy of early syn-rift tidal sediments: the Nukhul Formation, Suez Rift, Egypt. J Sediment Res 73:407–420

Carter R, Johnson D, Hooper K (1993) Episodic post-glacial sea-level rise and the sedimentary evolution of a tropical continental embayment (Cleveland Bay, Great Barrier Reef shelf, Australia). Aust J Earth Sci 40:229–255

Clifton HE (1983) Discrimination between subtidal and intertidal facies in Pleistocene deposits, Willapa Bay, Washington. J Sediment Petrol 53(2):353–369

Clough BF (1992) In: Clough BF (ed) Mangrove ecosystems in Australia, structure, function, and management. Australian National University Press, Canberra, pp 3–17

Clough BF, Andrews TJ, Cowan IR (1992) Physiological processes in mangroves. In: Clough BF (ed) Mangrove ecosystems in Australia, structure, function, and management. Australian National University Press, Canberra, pp 193–210

Curran HA (1985) The trace fossil assemblage of a Cretaceous nearshore environment: Englishtown formation of Delaware, U.S.A. In: Curran HA (ed) Biogenic structures: their use in interpreting depositional environments, vol 35. SEPM Special Publication, pp 261–276

Curran HA, Frey RW (1977) Pleistocene trace fossils from North Carolina (U.S.A.), and their Holocene analogues. In: Crimes TP, Harper JC (eds) Trace fossils 2. Geological Journal Special Issue No. 9. Seel House Press, Liverpool, pp 139–162

Dalrymple RW (1992) Tidal depositional systems. In: Walker RG, James NP (eds) Facies models: response to sea level change. Geological Association of Canada, St. John's, pp 195–218

Dalrymple RW (2006) Incised valleys in time and space; an introduction to the volume and an examination of the controls on valley formation and filling. Spec PublSoc Sediment Geol 85:5–12

Dalrymple RW, Choi K (2007) Morphologic and facies trends through the fluvial-marine transition in tide-dominated depositional systems; a schematic framework for environmental and sequence stratigraphic interpretation. Earth Sci Rev 81:135–174

Dalrymple RW, Boyd R, Zaitlin BA (1994) History of research, valley types and internal organization of incised-valley systems; introduction to the volume. In: Dalrymple RW, Boyd R, Zaitlin BA, Tulsa OK (eds) Incised-valley System; Origin and Sedimentary Sequences, vol 51. Special Publication SEPM (Society for Sedimentary Geology), pp 3–10

Dalrymple RW, Knight RJ, Zaitlin BA, Middleton GV (1990) Dynamics and facies model of a macrotidal sand-bar complex, Cobequid Bay -Salmon River estuary (Bay of Fundy). Sedimentology 37:577–612

Dalrymple RW, Zaitlin BA, Boyd R (1992) Estuarine facies models: conceptual basis and stratigraphic implications. J Sediment Petrol 62:1130–1146

De Gibert J, Netto R, Tognoli F, Grangeiro M (2006) Commensal worm traces and possible juvenile thalassinidean burrows associated with *Ophiomorpha nodosa*, Pleistocene, southern Brazil. Palaeogeogr Palaeoclimatol Palaeoecol 230:70–84

De Raaf JFM, Boersma JR (1971) Tidal deposits and their sedimentary structures (seven examples from western Europe). Geol Mijnbouw 50:479–501

Duke N (2006) Australia's mangroves: the authoritative guide to Australia's mangrove plants, vol 200. University of Queensland, Brisbane, pp 27–47

Dworschak P (1983) The biology of Upogebia pusilla (Petagna)(Decapoda, Thalassinidea) I. The burrows. PSZN I: Marine. Ecology 4:19–43

El Gezeery MN, Marzouk IM (1974) Miocene rock stratigraphy of Egypt. Egypt J Geol 18:1–59

Elliott T (1986) Siliciclastic shorelines. In: Reading HG (ed) Sedimentary environments and facies. Blackwell Scientific Publications, Oxford, pp 155–188

Ellison J (1988) Impacts of sediment burial on mangroves. Mar Pollut Bull 37:420–426

Erickson BR, Sanders AE (1991) Bioturbation structures in Pleistocene coastal plain sediments of South Carolina, North America. Sci Publ Sci Mus Minn 7:5–14

Fairbridge RW (1980) The estuary; its definition and geodynamic cycle. In: Olausson E, Cato I (eds) Chemistry and biogeochemistry of estuaries. Wiley, Chichester, pp 1–35

Falcon-Lang LH (1998) The impact of wildfire on an early carboniferous coastal environment, North Mayo, Ireland. Palaeogeogr Palaeoclimatol Palaeoecol 139:121–138

Finzel ES, Ridgway KD, Reifenstuhl RR, Blodgett RB, White JM, Decker PL (2009) Stratigraphic framework and estuarine depositional environments of the Miocene Bear Lake Formation, Bristol Bay Basin, Alaska: onshore equivalents to potential reservoir strata in a frontier gas-rich basin. AAPG Bull 93:379–405

Frey R, Howard J (1975) Endobenthic adaptations of juvenile thalassinidean shrimp. Bull Geol Soc Den 24:283–297

Frey RW, Howard JD, Pryor WA (1978) *Ophiommorpha*: its morphologic, taxonomic, and environmental significance. Paleogeogr Palaeoclimatol Palaeoecol 23:199–229

Friedman GM, Sanders, JE (1978) Principles of sedimentology. Wiley, New York, p 792

Furukawa K, Wolanski E (1996) Sedimentation in mangrove forests. Mangroves Salt Marshes 1:3–10

Furukawa K, Wolanski E, Mueller H. (1997) Currents and sediment transport in mangrove forests: Estuarine. Coast Shelf Sci 44:301–310

Galloway WE (1989) Genetic stratigraphic sequences in basin analysis: architecture and genesis of flooding surface bounded depositional units. AAPG Bull 73:125–142

Gingras M, George Pemberton S, Saunders T (2001) Bathymetry, sediment texture, and substrate cohesiveness; their impact on modern Glossifungites trace assemblages at Willapa Bay, Washington. Palaeogeogr Palaeoclimatol Palaeoecol 169:1–21

Gingras M, Hubbard S, Pemberton S, Saunders T (2000) The Significance of Pleistocene Psilonichnu. at Willapa Bay, Washington. Palaios 15:142

Gingras MK, Pemberton SG, Saunders TDA, Clifton HE (1999) The ichnology of modern and Pleistocene brackish-water deposits at Willapa Bay, Washington: variability in estuarine settings. Palaios 14:352–374

Gingras MK, Rasanen M, Ranzi A (2002) The significance of bioturbated inclined heterolithic stratification in the southern part of the Miocene Solimoes Formation, Rio Acre, Amazonia Brazil. Palaios 17:591–601

Giresse P, Lamboy M, Odin GS (1980) Evolution géométrique des supports de glauconitisation, reconstitution de leur paléo-environnement. Oceanologia Acta 3:251–260

Griffis R, Chavez F (1988) Effects of sediment type on burrows of Callianassa californiensis Dana and C. gigas Dana. J Exp Mar Biol Ecol 117:239–253

Hampson GJ, Procter EJ, Kelly C (2008) Controls on isolated shallow-marine sandstone deposition and shelf construction; Late Cretaceous Western Interior Seaway, northern Utah and Colorado, U.S.A. Spec Publ Soc Sediment Geol 90:355–389

Harris PT, Pattiaratchi CB, Cole AR, Keene JB (1992) Evolution of subtidal sandbanks in Moreton Bay, eastern Australia. Mar Geol 103:225–247

Harris M, Thayer P, Amidon M (1997) Sedimentology and depositional environments of middle Eocene terrigenous-carbonate strata, southeastern Atlantic Coastal Plain, USA. Sed Geol 108:141–161

Harris PT, Hughes MG, Baker EK, Dalrymple RW, Keene JB (2004) Sediment transport in distributary channels and its export to the pro-deltaic environment in a tidally-dominated delta: Fly River, Papua New Guinea. Cont Shelf Res 24:2431–2454

Heap AD, Bryce S, Ryan DA (2004) Facies evolution of Holocene estuaries and deltas; a large-sample statistical study from Australia. Sed Geol 168:1–17

Hesselbo SP, Huggett JM (2001) Glaucony in Ocean-Margin sequence stratigraphy (Oligocene–Pliocene, OffShore New Jercy, U.S.A.; ODP LEG 174A). J Sediment Res 71(4):599–607

Huggett JM, Gale AS (1997) Petrology and palaeoenvironmental significance of glaucony in the Eocene succession at Whitecliff Bay, Hampshire Basin, U.K. Geol Soc Lond 154:897–912

Ireland BJ, Curtis CD, Whiteman JA (1983) Compositional variation within some glauconites and illites and implications for their stability and origins. Sedimentology 30:769–786

Jackson C, Gawthorpe R, Carr I, Sharp I (2005) Normal faulting as a control on the stratigraphic development of shallow marine syn-rift sequences: the Nukhul and Lower Rudeis Formations, Hammam Faraun fault block, Suez Rift, Egypt: Sedimentology 52:313–338

Jaeger JM, Nittrouer CA (1995) Tidal controls on the formation of fine-scale sedimentary strata near the Amazon River mouth. Mar Geol 125:259–281

Johnson H, Levell B (1995) Sedimentology of a transgressive, estuarine sand complex: the Lower Cretaceous Woburn Sands (Lower Greensand), southern England: sedimentary facies analysis. Spec Publ Int Assoc Sedimentol 22:17–46

Kelly JC, Webb JA (1999) The genesis of glaucony in the Oligo-Miocene Torquay Group, southeastern Australia: petrographic and geochemical evidence. Sediment Geol 125:99–114

Kitamura A (1998) Glaucony and carbonate grains as indicators of the condensed section: Omma formation, Japan. Sed Geol 122:151–163

Kitazawa T (2007) Pleistocene macrotidal tide-dominated estuary-delta succession, along the Dong Nai River, southern Vietnam. Sed Geol 194:115–140

Klein G Dev (1970) Depositional and dispersal dynamics of intertidal sand bars. J Sediment Petrol 40:1095–1127

Kreisa R, Moila R (1986) Sigmoidal tidal bundles and other tide-generated sedimentary structures of the Curtis Formation, Utah. Bull Geol Soc Am 97:381

Leo RF, Barghoorn ES (1976) Silicification of wood: Botanical Museum Leaflets, vol 25. Harvard University, Cambridge, p 47

Lindholm R (1987) A practical approach to sedimentology. Allen & Unwin, London 276 p

Løseth TM, Ryseth AE, Young M (2009) Sedimentology and sequence stratigraphy of the Middle Jurassic Tarbert Formation, Oseberg South area (northern North Sea). Basin Res 21:597–619

Loutit TS, Hardenbol J, Vail PR, Baum GR (1988) Condensed sections: the key to age determination and correlation of continental margin sequences. In: Wilgus CK, Hastings BS, Kendall CGSC, Posamentier HW, Ross CA, Van Wagoner JC (eds) Sea-level changes: an integrated approach, Society of Economic Paleontologists and Mineralogists, Special Publication 42:183–213

MacEachern JA, Bann KL (2008) The role of ichnology in refining shallow marine facies models. In: Hampson G, Steel R, Burgess P, Dalrymple R (eds) Recent advances in models of siliciclastic shallow-marine stratigraphy, vol 90. SEPM (Society for Sedimentary Geology) Special Publication, pp 73–116

Maguregui J, Tyler N (1991) Evolution of middle Eocene tide-dominated deltaic sandstones, Lagunillas Field, Maracaibo Basin, western Venezuela. In: Miall AD, Tyler N (eds) The three-dimensional facies architecture of terrigenous clastic sediments, and its implications for hydrocarbon discovery and recovery, SEPM (Society for Sedimentary Geology), Concepts in Sedimentology and Paleontology 3:233–244

Martino RL, Curran HA (1990) Sedimentology, ichnology and paleoenvironments of the Upper Cretaceous Wenonah and Mt. Laurel Formation, New Jersey. J Sediment Petrol 60(1):125–144

McCracken SR, Compton, J., and Hicks, K., 1996, Sequence-stratigraphic significance of glaucony-rich lithofacies at Site 903. In: Mountain GS, Miller KG, Blum P, Poag CW, Twitchell DC (eds) Proceedings of the Ocean Drilling Program, Scientific Results, pp 171–187

Mccrimmon GG, Arnott RWC (2002) The Clearwater Formation, Cold Lake, Alberta: a world class hydrocarbon reservoir hosted in a complex succession of tide-dominated deltaic deposits. Bull Can Pet Geol 50:370–392

Mcgill JT (1958) Map of coastal landforms of the world. Geogr Rev 48:402–405

Middleton GV (1991) A short historical review or clastic tidal sedimentology. In: Smith DG, Reinson GE, Zaitlin BA, Rahmani RA (eds) Clastic tidal sedimentology. Can Soc Petrol Geol 16:ix–xv

Miller MF, Curran HA, Martino RL (1998) *Ophiomorpha* nodosa in estuarine sands of the lower Miocene Calvert Formation at the Pollack Farm Site, Delaware. In: Benson RN (ed) Geology and paleontology of the lower Miocene Pollack Farm Fossil Site, Delaware, Delaware Geological Survey Special Publication No. 21, pp 41–46

Milliman J (1972) Atlantic continental shelf and slope of the United States-petrology of the sand fraction of sediments, northern New Jersey to southern Florida. US Geol Surv 529-J:Jl–J40

Myrow PM (1995) Thalassinoides and the enigma of Early Paleozoic open-framework burrow systems. Palaios 10:58–74

Nio SD, Yang CS (1991a) Diagnostic attributes of clastic tidal deposits: a review. In: Smith DG, Reinson GE, Zaitlin BA, Rahmani RA (eds) Clastic tidal sedimentology. Canadian Society of Petroleum Geologists Memoir, Calgary, pp 3–28

Nio SD, Yang CS (1991b) Sea-level fluctuations and geometric variability of tide-dominated sandbodies. Sed Geol 70:161–193

Odin G (1988) Glaucony from the Gulf of Guinea. Green Marine Clays: oolitic ironstone facies, verdine facies, glaucony facies and celadonite-bearing facies; a comparative study, Amsterdam, Elsevier, Developments in Sedimentology. GS Odin 45:205–217

Odin G, Dodson M (1982) Zero isotopic age of glauconies. In: Odin GS (ed) Numerical dating in stratigraphy. Wiley, Chichester, pp 277–305

Odin GS, Fullagar PD (1988) Geological significance of the glaucony facies. In: Odin GS (ed) Green Marine Clays: oolitic ironstone facies, verdine facies, glaucony facies and celadonite-bearing facies; a comparative study. Elsevier, Developments in Sedimentology, Amsterdam, pp 295–332

Odin GS, Matter A (1981) De glauconiarum origine. Sedimentology 28:611–641

Olivero EB, Ponce JJ, Martinioni R (2008) Sedimentology and architecture of sharp-based tidal sandstones in the upper Marambio Group, Maastrichtian of Antarctica. Sediment Geol 210:11–26

Olsen T, Mellere D, Olsen T (1999) Facies architecture and geometry of landward-stepping shoreface tongues: the Upper Cretaceous Cliff House Sandstone (Mancos Canyon, south-west Colorado). Sedimentology 46:603–625

Pemberton S (1992) Applications of ichnology to petroleum exploration—a core workshop. Society of Economic Paleontologists and Mineralogists, p 429

Pemberton SG, Spila M, Pulham AJ, Saunders T, MacEachern JA, Robbins D, Sinclair IK (2001) Ichnology and sedimentology of shallow to marginal marine systems: Ben Nevis and Avalon Reservoirs Jeanne d'Arc Basin. Ottawa, Geological Association of Canada. Short Course Notes 15:343

Perry CT, Berkeley A, Smithers SG (2008) Microfacies characteristics of a tropical, mangrove-fringed shoreline, Cleveland Bay, queensland, Australia: sedimentary and taphonomic controls on mangrove Facies development. J Sediment Res 78:77–97

Pickett TE, Kraft JC, Smith K (1971) Cretaceous burrows—Chesapeake and Delaware Canal, Delaware. J Paleontol 45:209–211

Plaziat J (1974) Mollusc distribution and its value for recognition of ancient mangroves. In: International symposium on the biology and management of Mangroves, Honolulu, pp 456–465

Plaziat J (1995) Modern and fossil mangroves and mangals: their climatic and biogeographic variability. In: Bosence D, Allison P (eds) Marine palaeoenvironmental analysis from Fossils, Geological Society London, Special Publications, pp 73–96

Pontén A, Plink-Björklund P (2009) Process regime changes across a regressive to transgressive turnaround in a Shelf-Slope Basin, Eocene Central Basin of Spitsbergen. J Sediment Res 79:2–23

Porebski S (2000) Shelf-valley compound fill produced by fault subsidence and eustatic sea level changes, Eocene LaMeseta Formation, Seymour Island, Antarctica. Geology 28:147–150

Pritchard DW (1967) What is an estuary; physical viewpoint. In: Lauff GH (ed) Estuaries. AAAS Publication (American Association for the Advancement of Science), Washington, DC, pp 3–5

Ranger M, Pemberton S (1992) The Sedimentology and ichnology of estuarine point bars in the McMuarry Formation of the Athabasca Oil Sands Deposit, northeastern Alberta, Canada. Applications of Ichnology to Petroleum Exploration, vol 17. Pemberton, S. G., SEPM Core Workshop Notes, pp 401–421

Reading HG, Collinson JD (1996) Clastic coasts. In: Reading HG (ed) Sedimentary environments; processes, facies and stratigraphy. Blackwell Science, Oxford, pp 154–231

Reineck HE, Singh IB (1973) Depositional sedimentary environments with references to terrigenous clastics. Springer, New York, 431 p

Reineck HE, Wunderlich F (1968) Classification and origin of flaser and lenticular bedding. Sedimentology 11:99–104

Retallack GJ (1977) Triassic palaeosols in the upper Narrabeen Group of New South Wales. Part II. Classification and reconstruction. J Geol Soc Aust 24:19–35

Rice A, Chapman C (1971) Observations on the burrows and burrowing behaviour of two mud-dwelling decapod crustaceans, Nephrops norvegicus and Goneplax rhomboides. Marine Biol 10:330–342

Roberts EM (2007) Facies architecture and depositional environments of the Upper Cretaceous Kaiparowits Formation, southern Utah. Sediment Geol 197:207–233

Rogers K, Saintlan N, Cahoon D (2005) Surface elevation dynamics in a regenerating mangrove forest at Homebush Bay, Australia. Wetl Ecol Manage 13:87–598

Rossetti DF, Júnior AES (2004) Facies architecture in a tectonically influenced estuarine incised valley fill of Miocene age, northern Brazil. J S Am Earth Sci 17:267–284

Räsänen ME, Linna AM, Santos JCR, Negri FR (1995) Late Miocene tidal deposits in the Amazonian foreland basin. Science 269:386–390

Saenger P (1992) Morphological, anatomical, and reproductive adaptation of Australian mangroves. In: Clough BF (ed) Mangrove ecosystems in Australia, structure, function, and management. Australian National University Press, Canberra, pp 153–192

Said R (1962) The geology of Egypt. New York, Elsevier Publishing Company, Amsterdam 370 p

Scholle P, Ulmer-Scholle D (2003) A color guide to the petrography of carbonate rocks: grains, textures, porosity, diagenesis. Am Assoc Petrol Geol Bull Mem 77:474

Scurfield G, Segnit ER (1984) Petrification of wood by silica minerals. Sed Geol 39:149–167

Shanley KW, McCabe PJ, Hettinger RD (1992) Tidal influence in Cretaceous fluvial strata from Utah, U.S.A.: a key to sequence-stratigraphic interpretation. Sedimentology 39:905–930

Shanmugam G (1988) Origin, recognition, and importance of erosional unconformities in sedimentary basins. In: Kleinspehn KL, Poala C (eds) New perspectives in basin analysis. Springer, New York, pp 83–108

Shanmugam G (2003) Deep-marine tidal bottom currents and their reworked sands in modern and ancient submarine canyons. Mar Pet Geol 20:471–491

Shanmugan G, Poffenberger M, Alava JT (1998) Tide-dominated estuarine facies in the Hollin and Napo ("T" and "U") formations (Cretaceous), Sacha Field, Oriente Basin, Ecuador. In: Annual meeting expanded abstracts—American Association of Petroleum Geologists, vol 1998

Shanmugam G, Poffenberger M, Toro Alava J, Shanmugan G, Alava JT, Anonymous (2000) Tide-dominated estuarine facies in the Hollin and Napo ("T" and "U") Formations (Cretaceous), Sacha Field, Oriente Basin, Ecuador. AAPG Bull 84:652–682

Shinn E (1968) Burrowing in recent lime sediments of Florida and the Bahamas. J Paleontol 42:879–894

Smith DG (1988) Tidal bundles and mud couplets in the McMurray Formation, Northeastern Alberta, Canada. Can Soc Petrol Geol 36:216–219

Smith DG (1989) Comparative sedimentology of mesotidal (2 to 4 m) estuarine channel point bar deposits from modern examples and ancient Athabasca oil sands (Lower Cretaceous), McMurray Formation. In: Reinson GE (ed) Modern and ancient examples of clastic tidal deposits; a core and peel workshop. Canadian Society of Petroleum Geologists, Calgary, pp 60–65

Spalding M, Blasco F, Field C (1997) World Mangrove Atlas, Okinawa: Japan. The international society for Mangrove ecosystems, p 178

Stein CL (1982) Silica recrystallization in petrified wood. J Sediment Petrol 52(4):1277–1282

Stonecipher SA (1999) Genetic characteristics of glauconite and siderite; implications for the origin of ambiguous isolated marine sandbodies. In: Bergman KM, Sneddon JW (eds) Isolated shallow marine sand bodies; Sequence stratigraphic analysis and sedimentologic interpretation, vol 64. SEPM, Special Publication, pp 191–204

Strahler AN, Strahler AH (1974) Introduction to environmental science. Hamilton Publ. Co., Santa Barbara 633 p

Stratigraphic Sub-Committe of the National Committe Geological Sciences (1974) Miocene rock stratigraphy of Egypt. Egypt J Geol Cairo, National Information and Documentation Centre (NIDOC), 18:1–69 pp

Terwindt JHJ (1981) Origin and sequences of sedimentary structures in inshore mesotidal deposits of the North Sea. In: Nio SD, Shuttenhelm RTE, Van Weering Tj CE (eds) Holocene marine sedimentation in the North Sea Basin, International Association of Sedimentologists, Special Publication, pp 4–26

Thomas RG, Smith DG, Wood, JM, Visser J, Calverly-Range EA., Koster EH (1987) Inclined heterolithic stratification: terminology, description, interpretation and significance. Sediment Geol 53:123–179

Tucker M, Wright VP, Dickson JAD (1990) Carbonate sedimentology. Wiley-Blackwell, Oxford, 482 p

Visser MJ (1980) Neap-Spring cycles reflected in Holocene subtidal large-scale bedform deposits: a preliminary note. Geology 8:543–546

Waller T (1991) Evolutionary relationships among commercial scallops (Mollusca: Bivalvia: Pectinidae). Scallops Biol Ecol Aquacult 1:73

Walsha JP, Nittrouer CA (2004) Mangrove-bank sedimentation in a mesotidal environment with large sediment supply, Gulf Papua Marine Geol 208:225–248

Wells JT (1995) Tidal-dominated estuaries and tidal rivers. In: Perillo GME (ed) Geomorphology and sedimentology of estuaries. Series Development in Sedimentology, vol 53. Elsevier, Amsterdam, pp 179–205, 179–205, 471 p)

Willis BJ (2005) Deposits of tide-influenced river deltas. In: Giosan L, Bhattacharya JP (eds) River Deltas—concepts, models and examples: special publication, Society of Economic Paleontologists and Mineralogists, pp 87–129

Willis BJ, Bhattacharya JP, Gabel SL, White CD (1999) Architecture of a tide-influenced river delta in the Frontier Formation of central Wyoming, USA. Sedimentology 46:667–688

Willis BJ, Gabel S (2001) Sharp-based, tide-dominated deltas of the Sego Sandstone, Book Cliffs, USA. Sedimentology 48:479–506

Willis BJ, Gabel SL (2003) Formation of deep incisions into Tide-Dominated River Deltas: implications for the stratigraphy of the sego sandstone, Book Cliffs, Utah, U.S.A. J Sediment Res 73:246–263

Witzke BJ, Ludvigson GA, White TS Brenner RL (1999) Marine-influenced sedimentation in the Dakota Fm, Cretaceous (Albian—Cenomanian), central U.S.; Implications for sequence stratigraphy and paleogeography in the Western Interior. Geol Soc Am 31:425 (Abstracts with program)

Wonham J, Elliott T (1996) High-resolution sequence stratigraphy of a mid-Cretaceous estuarine complex: the Woburn Sands of the Leighton Buzzard area, southern England. Geol Soc Lond Spec Publ 103:41

Woodroffe C (1992) Mangrove sediments and geomorphology. In: Robertson A, Alongi D (eds) Tropical Mangrove ecosystems. American Geophysical Union, Coastal and Estuarine Studies, Washington, DC, pp 7–41

Woodroffe C, Bardsley K, Ward P, Hanley J (1988) Production of mangrove litter in a macrotidal embayment, Darwin Harbour, NT, Australia. Estuar Coast Shelf Sci 26:581–598

Woodroffe C, Grindrod J (1991) Mangrove biogeography: the role of Quaternary environmental and sea-level change. J Biogeogr 18:479–492

Yoshida S, Johnson HD, Pye K, Dixon RJ, Bann KL, Fielding CR, Maceachern JA, Tye SC (2004) Transgressive changes from tidal estuarine to marine embayment depositional systems: the Lower Cretaceous Woburn Sands of southern England and comparison with Holocene analogs differentiation of estuarine and offshore marine deposits using integrated ichnology and sedimentology; Permian Pebbley Beach Formation, Sydney Basin, Australia. AAPG Bull 88:1433–1460

Zaitlin BA, Dalrymple RW, Boyd R (1994) The stratigraphic organization of incised-valley systems associated with relative sea-level change. In: Dalrymple RW, Boyd R, Zaitlin BA (eds) Incised-valley systems: origin and sedimentary sequences, vol 51. SEPM Special Publication, pp 45–60

Chapter 4
Petrography of Carbonates (Microfacies Association)

Abstract The carbonate strata of the Moghra Formation represent a minor component of the studied rock succession. The carbonate facies are represented by fossiliferous limestones. Six distinct micro-facies (F) have been recognized: The most common micro-facies are mud-dominated packstone, quartz skeletal mud-dominated dolo-packstone, siliciclastic grain-dominated wackstones to siliciclastic grain-dominated packstones, siliciclastic grain-dominated packstones, quartz peloidal dolo-grainstone and quartz rich caliches or peletal fabric with caliches. These carbonate sediments that are represented by fossiliferous limestone are constructed exclusively of skeletal remains, notably bryozoans, bivalves, red algae, benthic foraminifers and barnacles. In addition, non-skeletal carbonate grains are represented by peloids and few aggregates. The depositional environments are interpreted as supratidal and intertidal to normal marine subtidal environments.

4.1 Introduction

The carbonate strata of the Moghra Formation represent a minor component of the studied rock sequence. They have been subjected to microfacies analysis using binocular and polarizing microscopes. The qualitative and quantitive analysis of the allochems of these rocks is based on the study of fifty thin sections. The limestone microfacies are differentiated and described based on the classifications of Dunham (1962) with the modifications of Embry and Klovan (1972) as well as the checklists and standard microfacies (SMF) types and facies zones (FZ) of Flügel (1982, 2004) and Wilson (1975). Dolomite bearing carbonates are subdivided based on the percentage of dolomite following the subdivision of Pettijon (1975): dolomitic limestone (10–50 %), limy dolostone (50–90 %) and dolostone (90–100 %).

S. M. Hassan, *Sequence Stratigraphy of the Lower Miocene Moghra Formation in the Qattara Depression, North Western Desert, Egypt*, SpringerBriefs in Earth Sciences, DOI: 10.1007/978-3-319-00330-6_4, © The Author(s) 2013

Table 4.1 Summary of the microfacies, defined in the carbonate strata of the Moghra Formation and their environmental interpretations

Facies Association	Facies	Microfacies	Depositional environment
Bioturbated-fossiliferous carbonate grain-dominated packstone facies association (FA6)	Fossiliferous limestone (F15)	Mud-dominated packstone	Normal marine subtidal environment
		Quartz-skeletal mud-dominated dolo-packstone	Open sea shelf and shelf lagoon with open circulation
		Siliciclastic grain-dominated wackestones to siliciclastic grain-dominated packstones	Shallow marine with moderate water circulation
		Siliciclastic grain-dominated packstone	Protected shallow marine environment with moderate water circulation
		Quartz peloidal dolo-grainstone	Protected inner-shelf
		Quartz rich caliches or peletal fabric with caliches	Supratidal and interdial

Six microfacies groups are recognized, each including a number of microfacies types that are more or less similar in their composition (Table 4.1).

4.2 Bioturbated-Fossiliferous Carbonate Grain-Dominated Packstone Facies Association (FA6)

4.2.1 Fossiliferous Limestone (F15)

This facies is limited to the northern and western part of the area in sections (5, 5′, 6, 7, 9, 10, 21 & 22). It accumulated maximum thicknesses in section 10 of about 9 m and consists of white color sandy fossiliferous limestone (Fig. 4.1a, b, c & d). It can be divided into six distinct micro-facies (F): The most common microfacies are mud dominated packstone, quartz skeletal mud-dominated dolo-packstone, siliclastic grain-dominated wackestones to siliciclastic grain-dominated packstones, siliciclastic grain-dominated packstone, quartz-peliodal dolo-grainstone and quartz-rich caliches or pelletal fabric with caliches, which correspond to standard microfacies 10, 10, 9, 16, 16 and 17 (Fig. 4.2), of Wilson (1975).

Fig. 4.1 Photographs of common lithofacies in Bioturbated-grain-dominated packstone facies association (FA6). (**a** & **b**) Fossiliferous limestone. (**c**) Abundant rip-up clasts in the base of this facies (*yellow arrows*). (**d**) Close-up view of echinoid shell in Fossiliferous limestone

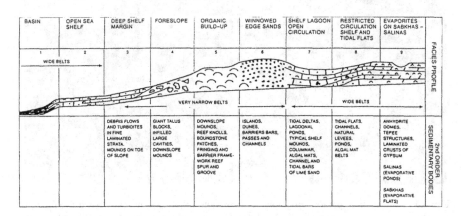

Fig. 4.2 Facies belt model modified from Wilson (1975)

4.2.1.1 Mud-Dominated Packstone

Description

This facies is recorded from section 9 (bed no. 9-6). This facies occurs within an 8 m thick limestone unit and this could be representing the Marmarica Formation or the top upper most of Moghra Formation.

This facies contains bryozoans, echinoderms, and red algae with minor pelecypods, gastropods, foraminifera, serpulids and abraded skeletal debris. Bryozoan fragments are preserved (Fig. 4.3a, b & c). A variety of common bryozoan growth habits are observed in the facies including: sheets, branches (ramose), and fenestrate forms. These specimens are represented dominantly by cheilostome and cyclostome bryozoans. The echinoid material is distinguished by its unrecrystallized, originally calcitic, prismatic wall structure. In addition, cross section of an echinoid spine showing single-crystal optical behavior (unit extinction). The red

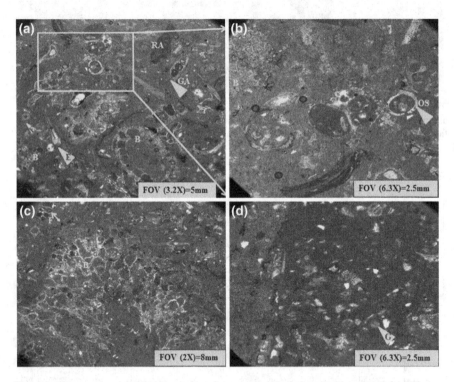

Fig. 4.3 Thin section photomicrographs of common lithofacies in bioturbated-fossiliferous sandstone facies association (FA5). Mud Dominated Packstone. (**a**) A longitudinal section through an echinoid spine and oblique section through the bryozoa. A fragmented gastropod that still is recognizable by shape. Although some internal layers are micrtizied but still visible, most of this originally aragonitic shell was dissolved and the mold was later filled with sparry calcite. (**b**) close view for photo A. (**c**) The shell at the *left top* is cut through large bryozoan colony B showing typical shape and structure. (**d**) Transition from micrite (*right*) to microspar, siderite and ferron dolomite (*left*). Note the boundary at the *left* from the photo

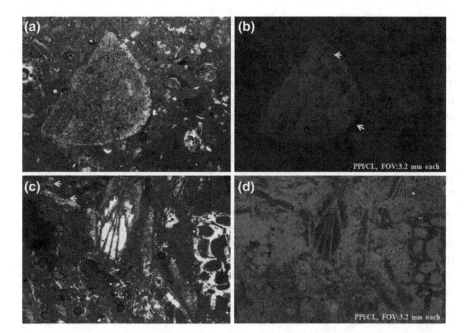

Fig. 4.4 Cathodoluminescence photomicrographs of common lithofacies in facies association fossiliferous limestone FA6. (**a & b**) This pair of photographs demonstrates the contrast between normal illumination and cathodoluminescence. Note that the echinoid shell is distinguished by bright luminescence and indicates conversion of high-Mg calcite to low-Mg calcite. (**c & d**) This pair of photographs demonstrates the contrast between normal illumination and cathodoluminescence. Note the bryozoans fragment was partially aragonite and dissolved and distinguished by non-luminescence

algae are well preserved here and it is highly micrtized. The oyster fragments from the calcitic bivalve. The gastropods also identified and it has been replaced with sparry calcite. Besides, A fragmented, abraded, and neomorphosed gastropod shell that still is recognizable by shape (another specimen is visible in the lower right). Although some organic-rich internal layers are still visible, most of this originally aragonitic shell was dissolved and the mold was later filled with sparry calcite. Few grains from forminifera have found and represented by a biserial foraminifer is a tangential cut through a different specimen of the same species. Moreover, few organisms are agglutinated wall structure of this textulariid foraminfera and the interagranular sparite fill the chamber of the foraminifera. Beside, all these bioclasts we identified few fragments of serpulid tube which they can be identified by the cylindrical, tube like shell form and the microstructure. The glauconite grains are found but less dominated. The matrix is dominated from micrite. In addition, the transitions from micrite to microspar (Fig. 4.3d). Moreover, other detrital mineral from quartz is present.

Cathodoluminescent in this faciesis showing dolomitize for outer edges for the echinoid plate (Fig. 4.4a & b). The bright luminescence, and dull luminescence have distinguished. The bright luminescences from the calcite cement and the zones with little or no luminescence tend to have elevated Fe^{2+} contents and depleted in Mn^{2+}. In addition, the bryozoans shell fragment show two stages of diagenesis. The first the calcite cement that distinguished by bright luminescence. The second is the few chambers from the shell dissolved and it is identified with no luminescence (Fig. 4.4c & d). Furthermore, we have distinguished few molluscan shell fragments have replaced by dolomite rhomb and give dull to no luminescence. Multiple compositional zones of alternating bright and dull luminescence in the dolomite has not seen clear.

Interpretation

The Mud Dominated Packstoneis correlated with the standard microfacies type **SMF** 10. (Bioclastic packstones and grainstones with coated and abraded skeletal grains) of Wilson (1975) who assigned them to be occurred in shelf lagoon with open circulation (FZ 7) and open sea shelf (FZ 2). The presence of a diverse marine biota, including echinoids, bryozoans, mollusca, and foraminiferas, is indicative of a normalmarine subtidal environment. The abundant mud, peloids, and bioturbation suggest a moderate to low energy setting.

4.2.1.2 Quartz-Skeletal Mud-Dominated Dolo-Packstone

Description

This facies is recorded from section 9 (bed no. 9-4-2). This bed consists of 6 m yellow fine grained siliciclastic-rich dolo-packstone. This facies is characterized by bioturbation and karstic weathering. In addition, it contains shell fragments from pelecypods and pectin.

The main skeletal allochems in this facies are echinoderms, and bivalves. Problematic grain could be barnacle. Foraminifera, peloid and intraclast are preserved. The echinoid present The bivalve shells were originally of aragonitic minerology, and the aragonite dissolved out early during diagenesis, and leaving a mold which later filled with and chalcedony. The pores were filled with micritic material that contrasts with the optically clear calcite of the echinoid (Fig. 4.5a & b).The foraminifera are dominated by miliolid foraminifers The original walls were high-Mg calcite, although some of the champers may have been partially dissolved. Few grains from peloid and intraclasts are found (Fig. 4.5c). The intraclast distinguished by the cluster or aggregate of other grains are held together by micrite. The calcite cement is extensively replaced by siderite or ferron dolomite(Fig. 4.5d). Dolomite crystals are distinct. Although the crystal fabric is tightly interlocking, the rhombic shape of the dolomite crystals is clearly outlined by the zones. Moreover, other detrital component is represent by quartz in which dominated in silt size.

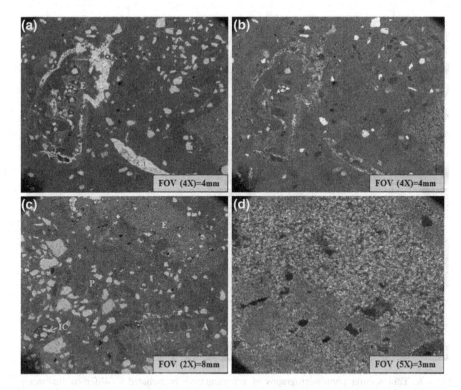

Fig. 4.5 Thin section photomicrographs of common lithofacies in bioturbated-fossiliferous limestone facies association (FA6). Quartz Skeletal Mud Dominated Dolo-Packstone. (**a** & **b**) An unusual of echinoid plate a strongly ornamented and very porous. Here the pores were filled with micritic material that contrasts with the optically clear calcite of the echinoid PPL, XPL. (**c**) Echinoid, transverse cross section through bryozoans fragment is distinguished PPL. Note the intraclast and peloid grains. (**d**) A dolomitization front in a micritic matrix—dolomitization is complete on the *right* side and sparse on the left XPL

Interpretation

This microfaciesis correlated with the standard microfacies type **SMF** 10, Bioclastic packstones and grainstones with coated and abraded skeletal grains. It is worn and coated bioclasts deposited within a fine-grained matrix. It is occurred in open sea shelf (FZ 2) and Shelf lagoon with open circulation (FZ 7), Wilson (1975) and Flügel (1982, 2004).

4.2.1.3 Siliciclastic Grain-Dominated Wackstones to Siliciclastic Grain-Dominated Packstones

Description

This facies is recorded from Section 22 (beds no. 22-19). This facies consists of 5 m of sandy limestone. The base is marked by lag deposits mainly from flint and

Fig. 4.6 Thin section photomicrographs of representative bioturbated-fossiliferous limestone facies association (FA6). Siliclastic Grain-Dominated Wackstones to Packstones. (**a** & **b**) Echinoid fragment E, with characteristic single-crystal or unit extinction and uniform"honey-comb" microtecture (small pores filled with micrite) PPL, XPL. (**c** & **d**) Specimen from bryoazoan B (*yellow arrow*) with thin zooecial wall an elongated zooecia PPl, XPL

few mud clasts. In addition, this facies is highly bioturbated with small burrows. Moreover, the sedimentary structure is represented by cross-bedded with dominant direction NE (25°, 45°, 33°).

Allochems characteristic of this facies include echinoids, bryozoa, and minor mollusk shell fragments. The echinoderms are represented by a large echinoiderm fragment with characteristic single or unit extinction and "honeycomb" microtexture (small pores filled with micrite) (Fig. 4.6a & b). The irregular shape and lack of central canal help to distinguish it from a columnal crinoid. Echinoid spines were identified. The transverse cross section through bryozoans fragment is distinguished (Fig. 4.6c & d). Original inter-particle pore space is usually filled with calcite cement and gypsum. Rare mollusk shells fragments (bivalves), which were originally composed of aragonite, were dissolved and replaced with sparry calcite or few grains left as skeletal molds. In addition, the wall of these shells micritzed. The matrix of micritic texture and is dolomitized. The top most part of this facies became sandier and contains echinoid shells.

Interpretation

This microfacies is correlated with the standard microfacies type SMF 9 (Strongly burrowed bioclastic wackestone) that occurred in shallow lagoon with open circulation at or just below the fair-weather wave base (FZ 7). In addition, it interprets as shallow shelf near siliciclastic source. This criteria will be known from dominant of micrite with common to abundant fragmented fossils (bivalves, gastropods, and echinoderms) that are jumbled through burrowing and the bioclasts often micritized, Wilson (1975) and Flügel (1982, 2004).

4.2.1.4 Siliciclastic Grain-Dominated Packstone

Description

This facies is recorded from section 21 (bed no. 21-7-2) and consists of 70 cm cross bedded lenticular beds. This bed entirely characterized by mud drape. This facies represented also in section 25 (1.5 m, 25-8) by sandy dolo-packstone and consists of yellowish brown hard, massive, moderately bioturbated fossiliferous limestone. This facies starts by a pebbly layer with elongated and oriented mudclasts. This facies contains abundant pelecypods. The base of this facies is also distinguished by thick horizontal Thalassiniodes networks. The external geometry is swalley.

The main skeletal allochems in this facies are ostracodes, echinoid and few grains from crinoids plated with molluscan shell are preserved too. The ostracodes make up a significant portion of the total deposit of this facies. The complete ostracode shows overlap of valves and a geopetal internal sediment fill (Fig. 4.7a & b). A fish-hook-like termination of a single ostracode valve and these terminations are distinctive and, in combination with carapace size, structure, and wall morphology helps to reliably identify ostracode (Fig. 4.7a). In addition, the presence of such overlapping margins and the absence of interlocked hinged terminations help us to distinguish ostracodes from small bivalves. Few grains from this ostracod has intraparticle pore that plugged with coarse, cavity-filling sparry calcite (Fig. 4.7b). The crystal sizes increase toward the center of the cavity and dominated in this case by just a few very large central crystals (see, Scholle and Ulmer-Scholle 2003). Furthermore, a relatively thick-walled, articulate ostracode shells is distinguished. A transverse section through a single echinoid spine is showing the characteristic lobate outline and flower-like structure. The structure of the spine is distinguished because of micritic infill. A crinoids plate has found and showed the unit extinction (single crystal extinction). The traces of pore structure, and the axial canal is characterized for this plate (Fig. 4.7c & d). The grain has been substantially altered by cementation within pores, by organic boring (and filling of those borings with micrite). The absence of syntaxial overgrowths also is a result of the presence of extensive micrite matrix around the grains (Scholle and Ulmer-Scholle 2003). The bivalves shells are identified with micritic envelop, and the shells were dissolved and the molds were later filled with sparry calcite (LMC) (Fig. 4.7d).

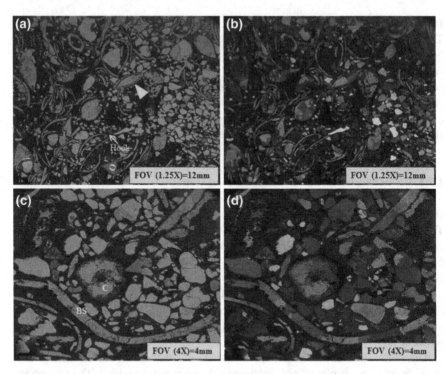

Fig. 4.7 Thin section photomicrographs of representative bioturbated-fossiliferous sandstone facies association (FA5). Siliciclastic grain dominated-packstone. (**a** & **b**) Ostracods are represented by disarticulated valves (*blue arrows*, see the hook with *yellow arrow*) and complete tests (*yellow arrows*) PPL, XPL. The bivalve shells (BS) are replaced by coarse calcite. (**c** & **d**) The bivalve shells (BS) were dissolved and the molds were later filled with sparry calcite and a crinoid fragment with single-crystal structure, a clearly displayed lumen (central canal) the pores at their margin are completely filled with micritic carbonate giving the grains a characteristic "dusty" appearance (*red arrow*) PPL, XPL

Gluaconite percentage is less than 5. These glauconite grains have a characteristic pale green to yellowish color; some are internally featureless, whereas others show cerebroid or lobate structures (according to Scholle and Ulmer-Scholle 2003). Some glauconite grains have been replaced by calcite and silica (Fig. 4.8a & b). In addition, few grains showed the internal ghost from the original source for the glauconite but we can't identify it. The feldspar identified here and form 1 %. Authigenic feldspars are present. However, the feldspar replacements can be easily confused with megaquartz replacements due to their similar birefringence, but it can be differentiated based on euhedral crystal shapes and, the presence of twinning. Quartz as detrital mineral represent as sand to silt size. The quartz grains are subrounded to well rounded, straight extension with little grain wavy extension.

Interpretation
This microfacies is correlated with the standard microfacies type **SMF** 16 Non-laminated peloidal grainstone and packstone and laminated peloidal. It is characterized by ostracods associated with benthic foraminifera. SMF 16- Non-laminated is common in shallow platform interiors comprising protected shallow-marine environments with moderate water circulation (FZ 8), Wilson (1975) and Flügel (1982, 2004).

4.2.1.5 Quartz Peloidal Dolo-Grainstone

Description
This facies is recorded from section 10 (bed no. 10-9-2) and consists of 9 m white color limestone

This facies contains foraminifera, bivalves, echinoids, and intraclasts. This facies contains a large proportion of peloids (Fig. 4.8c & d). In addition, highly

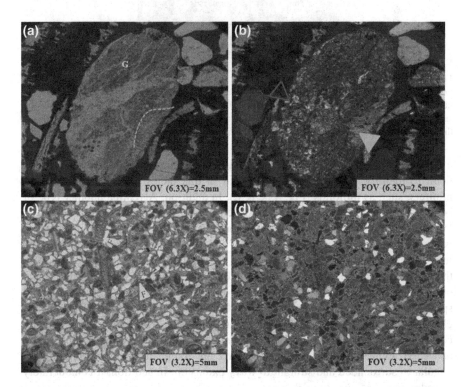

Fig. 4.8 Thin section photomicrographs of common lithofacies in bioturbated-fossiliferous limestone facies association (FA6). Quartz Peloidal Dolo-Grainstone. (**a** & **b**) Large Glaucinte grain and it is replacement by calcite (*orange arrow*) cement and silica (chalcedony quartz, *red arrow*) PPl, XPL. (**c** & **d**) Show miliolids, a type of foraminifera with many chambers. The matrix is micrite with many fragmented bioclasts, PPL, XPL

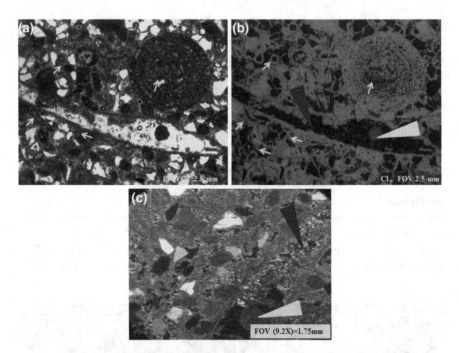

Fig. 4.9 Thin section photomicrographs of common lithofacies in facies association FA6, Fossiliferous limestone. (**a & b**) This pair of photographs demonstrates the contrast between normal illumination and cathodoluminescence. Note that the bivalve shell replaced partially by silica (chert and chalcedony) and it distinguish by non-luminescence. (**c**) Close view under shows the moluscan fragment replaced partially with chalcedony

micritized bioclasts often exhibit a peloid-like appearance and are therefore difficult to identify. The foraminiferas found in this facies include benthics foraminifers, such as miliolids (Fig. 4.8c & d). In addition, Alveolina another type of benthic foraminifera with chambers divided by septulae into numerous chamberlets arranged in one or more rows is observed (Fig. 4.9a). Moreover, few organisms are agglutinated wall structure of this textulariid foraminfera and the interagranular sparite fill the chamber of the foraminifera. The molluscan with thick micritic wall and dissolved and filled with silica (chert and chalcedony) and other micrtized. Echinoid fragments are identified and represented as echinoid spine. In addition, the large grain benthic foraminfera wall has micritized and the chambers itself fill with silica. Few grains from intraclasts are found. Sparry calcite cement is dominant. In addition, the sparry cement replaced by chalcedony quartz and chert. Other detrital grains from quartz are found. It is moncrystaline, subangular to surrounded and straight to wavy extension. In addition, it is moderately sorted.

Cathodoluminescent in this facies is showing silica replacement of molluscan fragment and vug-filling dolomite cements and calcite. It shows the preservation of some of the original carbonate rock fabric within the quartz crystals (no luminescence). The preserved fabric remains visible mainly through the presence of undigested remnants of carbonate as inclusions within the silica (Fig. 4.9a & b). The bright luminescence has seen from the calcite cement and reflects temporal geochemical changes allowing variations incorporation of Mn^{2+} (a CL-exciting ion) and Fe^{2+} (a CL-quenching ion). The molluscan shell dissolved and it is identified with no luminescence duo to the silica replacement (Fig. 4.9b & c). We have distinguished the multiple compositional zones of alternating bright and dull luminescence in the dolomite.

Interpretation

This microfacies is correlated with the standard microfacies type **SMF** 16- Non–laminated (equivalent to the grainstone with pellets in Wilson (1975) and Flügel (1982) and occurred in (FZ7) and (FZ8) which it is common in shallow platform interiors comprising protected shallow-marine environments with moderate water circulation (FZ 8) Wilson (1975) and Flugel (2004).

The abundance of miliolids is additionally indicative of a protected inner-shelf environment (Hallock and Glenn 1986). This facies is inferred to have deposited in a relatively protected inner-shelf environment. The abundance of miliolids foraminifera indicates water-depths greater than 20 m (Montaggioni 1981) cit in (Fournier et al. 2004); however, many miliolids facies much shallower.

4.2.1.6 Quartz Rich Caliches or Peletal Fabric with Caliches

Description

This facies is recorded from section 5 (bed no. 5-3-16) and consists of 1 m hard, compacted white color limestone with shell fragments. It is distinguished by highly karstified. In addition, it based by burrows that could be belong to the bed below it.

The caliches facies mainly consists of clotted texture from soil fabric coated grains and few grains from pellets. In this caliche we can note the irregularly shaped coated grains (termed soil pisoids or pisoliths, Fig. 4.10a & b) and abundant inclusions of detrital terrigenous silt and sand that are surrounded by pedogenic carbonate. These terrigenous components are concentrated during the dissolution process that characterizes long-term exposure surfaces (according to Scholle and Ulmer-Scholle 2003). In addition, few pellets of unknown origin, but it could be fecal origin. Quartz is medium-coarse sand. Few grains from peloid and intraclasts are found. The intraclast could be "grapestone" intraclast (Fig. 4.10c & d) and distinguished by the cluster or aggregate of other grains are held together by micritic along with microbial, and other encrustations (see Scholle and Ulmer-Scholle 2003). A transverse section through a single, originally aragonitic, gastropod is identified. All trace of original wall structure has been micritized and the mold itself filled by calcite. In addition, a thick band of radiaxial-fibrous calcite cement lines all former pores (Fig. 4.10c & d). The matrix from micrite and few

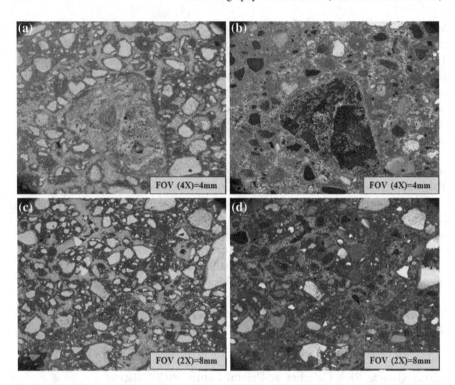

Fig. 4.10 Thin section photomicrographs in of common lithofacies in bioturbated-fossiliferous limestone facies association (FA6). Quartz Rich Calichesor Peletal- Fabric with Caliches. (**a & b**) Note the irregular, highly asymmetrical coated grains (could be pisoids/pisoliths) and abundant inclusions of detrital terrigenous sand and silt PPL, XPL. (**c & d**) A thick band of radiaxial-fibrous calcite cement lines all former pores (*red arrow*) PPL, XPL. Note the intraclast could be "grapestone" (*yellow arrows*), PPL, XPL

patches from sparry calcite cement. In addition, cement form of thin isopachous calcite rims is identified.

Interpretation
This facies belong to Standard Facies Zones FZ 10 of the modified Wilson model describing a rimmed carbonate platform, paleokarst, caliche and other terrestrial and terrestrial marine Settings (Humid and arid often subaerially exposed, meteorically influenced limestones). It is setting in subaerial or subaquatic, formed under meteoric-vadose and marine-vadose conditions. Abundant inkarst settings and pedogenic carbonates (continental andnear-coast areas), and supratidal and intertidal environments (Flügel 2004).

4.2.1.7 Interpretation (F15)

Fossiliferous limestone is interpreted to be the most seaward facies in the Moghra formation. The diverse fauna of facies suggests deposition in clear, well-oxygenated, open-marine water of normal salinity on the inner to middle shelf (~ 30 m depth) with periods of marginal marine, beach, and deltaic influence (Harris et al. 1997). The presence of whole, unabraded foraminifera, ostracodes, and gastropods with relatively unworn echinoderms, bryozoans, and pelecypods suggests a low- to moderate-energy marine shelf environment with open circulation. Nevertheless, the presence of broken, abraded, and rounded echinoderms, pelecypods, and bryozoans, intermixed or interbedded with unabraded allochems, indicates alternating energy regimes ranging from low to high (Flügel 1982).

According to the previous ternary diagram (Fig. 3.19) modified from Harris et al. (1997) (after Lindholm 1987) illustrating the proposed energy regimes and interpreted depositional environments for the fossiliferous limestone as follow:

The quartz-rich wackestone, and packstone lithofacies (sC) represent inner to middle shelf environments of low to moderate energy with possible deltaic or riverine influence supported by the presence of cellophane, organic debris, and subangular, fine- to medium-grained terrigenous quartz. The presence of glauconite along with the faunal assemblage of pelecypods, bryozoans, echinoderms, foraminifers, and other skeletal debris indicates a marine depositional environment of normal salinity that probably accumulated on the inner to middle shelf (Milliman 1972). The broken and rounded nature of the gravel-size pelecypods and bryozoans intermixed with sand size allochems is indicative of bottom transport by strong currents or storm-generated waves.

The packstone lithofacies (C) represent lower-energy environments of the middle shelf. The lack of terrigenous quartz and the abundance of mud also suggests quiet water deposition below normal wave base on the inner to middle shelf. The presence of glauconite and the dominance of echinoderms, bryozoans, foraminifers, and pelecypods, with lesser amounts of gastropods and ostracodes support a subtidal middle shelf setting.

The grainstones, which occur as thick facies, probably represent higher-energy conditions, and likely coincide with periods of shallower water depth less than 15 m deep (Thayer et al. 1993). The grainstones are interpreted to represent isolated shoals or storm surge deposits within the nearshore to inner shelf environment (Harris et al. 1997).

The grain dominated wackestone/packstone has a diverse fauna, including filter feeders such as bryozoa and crinoids, and indicates normal marine conditions (Lemone et al. 1971; Lemone et al. 1975). The predominance of micrite in most types of limestone reflects quiet-water deposition (Mack and James 1986).

In addition, echinoids are primarily restricted to waters of normal marine salinity. A few echinoids can tolerate transitional marine conditions (20–30 %), but none are known from brackish or freshwater settings (Durham 1966; Heckel 1972) cited in (Martino and Curran 1990).

Glauconite is usually regarded as an indicator of marine environment, relatively shallow deposition and slow sedimentation (Scholle and Ulmer-Scholle 2003).

All the above criteria support open-marine water of normal salinity on the inner to middle shelf for this facies association.

4.2.2 Discussion

There are many problems in the interpretation of the carbonate microfacies under the microscope. There are two factors which may play a role in the distribution of sediment components inside the Moghra Formation, including the transport of components into other environmental settings or the mixing of the sediment through bioturbation. These may explain the apparent inconsistency between the high percentage of filter feeding organisms (suggesting strong current activity), and the high amount of mud in most facies. In addition to these factors there are many erosion surfaces could be concerned; however, it will be auto-cyclic or allo-cyclic. Moreover, the vertebrate bone fragments have complicated the differentiation along with invertebrates. Although we have these problems to interpret the moicrofacies in Moghra Formation, we have to come up with conclusion that can be associated with these mircofacies.

4.2.3 Diagenesis

The diagenetic features observed in the thin section from the Moghra Formation are interpreted to have developed primarily in marine and freshwater phreatic environments. Diagenetic events affected the wackestones, packstones, and grainstones.

Marine phreatic events are interpreted to include: (1) micritization of allochems, primarily foraminifers and mollusks, resulting in formation of micrite envelopes and amorphous grains ('pellets') of micrite. In addition, micritization has preferentially affected red algae and porcelaneous benthonic foraminifera. Highly micritized bioclasts often exhibit a peloid-like appearance and are therefore difficult to identify; (2) micritization of the foraminifers and mollusks along with precipitation of glauconite cements is believed to have occurred shortly after deposition; (3) rare isopachous Mg-calcite cement rimming skeletal allochems; and (4) precipitation of glauconitic cements.

The following diagenetic events occurred in the, fresh-water phreatic environment probably during the sea level fall (1) conversion of high-Mg calcite allochems (primarily echinoderms) to low-Mg calcite. (2) formation of syntaxial low-Mg calcite overgrowths on echinoderm fragments. (3) dissolved of aragonitic allochems, primarily mollusks, and formation of moldic porosity. (4) precipitation of rare pore-reducing or pore-filling low-Mg calcite spar within intraparticle pores in foraminifers, ostracodes, and bryozoans (5) partial replacement of Mg-calcite molluskan shells and filling of borings within the shells by spherulitic chalcedony. (6) precipitation of isopachous low-Mg calcite cement on quartz and glauconite grains. (7) neomorphism of micrite matrix to microspar.

4.3 Carbonate Lithofacies as Paleolatitude Indicator Problem Within Moghra Formation

Has the paleolatitude of the carbonate lithofacies of Moghra Formation is related to subtropical or temperate water carbonate (Non-Tropical)?

Some ancient carbonate sequences are not easily correlated with the well-known shallow tropical-subtropical modern carbonates. Many of these ancient carbonates have common features: (1) abundance of mollusks, benthic foraminifers, encrusting coralline algae, bryozoans, echinoids and barnacles: and (2) lack of hermatypic corals and characteristic non-skeletal carbonate grains such as ooids, botryoidal grains or grapestones. This assemblage can be considered similar in composition to the foramol association of Lees and Buller (1972) and is very abundant in the modern temperate to subtropical open shelves of Mediterranean Sea (Carannante et al. 1981; Simone and Carannante 1985). As a consequence, this similarity raises the risk of misinterpreting all such ancient carbonate sequences as characteristic of the temperate climatic zone (Carannante et al. 1988).

However, the foramol-type sediments are very abundant on temperate shelves, but they also are present in shallow tropical or subtropical waters where reef corals and calcareous green algae are not developed because of particular environmental or developed because of particular environmental or ecological conditions (Carannante et al. 1988). This lithofacies could be named also as **molechfor** according to (Carannante et al. 1988), in which echinoids, barnacles, serpulids and bryozoans may associate with mollusks and arenaceous foraminifers.

In addition, it is well known, as pointed out by Lees (1974, 1975) and Lees and Buller (1972), that foramol-type sediments also may be present in tropical to subtropical latitudes. The schematic Fig. 4.11 it can explain that the **foramol** association extends well into the tropics; as a result, the belt from the equator to 30° latitude, water temperatures tend to be higher on the western than on the eastern sides of the oceans. Therefore, it is significant, however, that although these temperature thresholds broadly account for the known distribution of the skeletal associations, important anomalies remain. Moreover, Lees (1975) has shown for the modern situation how carbonates of cool-water aspect may extend into warmer waters of low latitudes and "replace" tropical carbonates as a consequence of dilution of normal-salinity sea water by fresh water (see Fig. 4.12, cit in Nelson 1988). On the other hand, we have to think about the controlling factors that can affect the foramol lihofacies development. For example, the interaction of

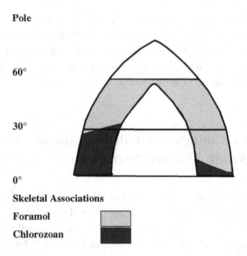

Fig. 4.11 Modified from Lees and Buller (1972)

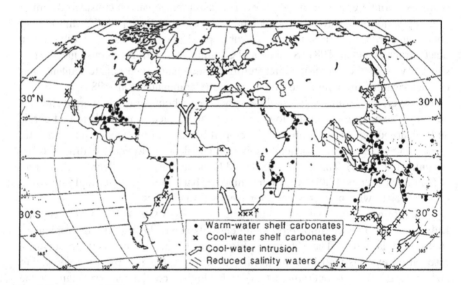

Fig. 4.12 Nelson (1988)

salinity-temperature was regarded by Lees and Buller (1972) and Lees (1973, 1974, 1975) as the main controlling factor in the development this foramol lithofacies. In addition, Nutrient concentrations have been proposed by Hallock and Schlager (1986) as the dominant factor limiting coral reef development (Carannante et al. 1988).

References

Carannante G, Simone L, Barbera C (1981) Calcari a briozoi e litotamni of Southern Apennines; Miocene analogs of recent Mediterranean rhodolitic sediments. International Association of Sedimentologists abstracts, 2nd European regional meeting. International Association Sedimentology, Oxford, p 17–20

Carannante G, Esteban M, Milliman JD, Simone L (1988) Carbonate lithofacies as paleolatitude indicators; problems and limitations. Sed Geol 60:333–346

Dunham RJ (1962) Classification of carbonate rocks according to depositional texture. In: Ham WE (ed) Classifications of carbonate rocks—a symposium: AAPG Memoir 1, pp 108–121

Durham JWAO (1966) Echinoids, ecology and paleoecology. In: Moore RC (ed) Treatise of invertebrate paleontology, Part U, Echinodermata 3, Asterozoa-Echinoizoa. University of Kansas Press, Geological Society of America, Lawernce, pp U257–U265

Embry AF, Kolvan JE (1972) Absolute water depths limits of Late Devonian paleoecological zones. Geol Rdsch 61:672–686

Flügel E (1982) Microfacies analysis of limestones. Springer, New York 633

Flügel E (2004) Microfacies of carbonate rocks. Analysis, interpretation and application. Springer, Berlin, p 976

Fournier F, Montaggioni L, Borgomano J (2004) Paleoenvironments and high-frequency cyclicity from Cenozoic South-East Asian shallow-water carbonates: a case study from the Oligo-Miocene buildups of Malampaya (Offshore Palawan, Philippines). Mar Pet Geol 21:1–21

Hallock P, Glenn EC (1986) Larger foraminifera: a tool for palaeoenvironmental analysis of Cenozoic carbonate depositional facies. Palaios 1:44–64

Hallock P, Schlager W (1986) Nutrient excess and the demise of coral reefs and carbonate platforms. Palaios 1:389–398

Harris M, Thayer P, Amidon M (1997) Sedimentology and depositional environments of middle Eocene terrigenous-carbonate strata, southeastern Atlantic Coastal Plain, USA. Sed Geol 108:141–161

Heckel PH (1972) Recognition of ancient shallow marine environments. In: Rigby JK, Hamblin WK (eds) Recognition of ancient sedimentary environments, SEPM(Society for Sedimentary Geology) Special Publication, pp 226–286

Lees A (1973) Les dépots carbonatés de plate-forme: Centre de Recherches de Pau (Societe Nationale des Petroles d'Aquitaine), Bulletin., v 7, p 177–192

Lees A (1974) Contrasts between recent warm-and cold-water carbonates: significance in the interpretation of ancient limestones. Annales de la Societé Géologique de Belgique, vol 97, pp 159–161

Lees A (1975) Possible influence of salinity and temperature on modern shelf carbonate sedimentation. Mar Geol 19:159–198

Lees A, Buller AT (1972) Modern temperate-water and warm-water shelf carbonate sediments contrasted. Mar Geol 13:M67–M73

Lemone DV, Klement KW, King WE (1971) Abo-Hueco facies of the Upper Wolfcamp Hueco formation of the southeastern Robledo Mountains, Dona Aria County, New Mexico. In: Cys JM (ed) Robledo Mountains, New Mexico, Franklin Mountains, Texas, Soc. Econ. Paleontologists and mineralogists–permian basin section field conference on guidebook, pp 137–172

Lemone DV, Szmpson RD, Klement KW (1975) Wolf-campion upper Hueco formation of the Robledo Mountains, Dona Ana County, New Mexico. In: Seager WR, Clemons RE, Callender JF (eds) Guidebook of the Los Cruces Country, 26th field conference, Geological Society, New Mexico, pp 119–121

Lindholm R (1987) A practical approach to sedimentology. Allen & Unwin, London, p 276

Mack G, James W (1986) Cyclic sedimentation in the mixed siliciclastic-carbonate Abo-Hueco transitional zone (Lower Permian), southwestern New Mexico. J Sediment Petrol 56:635–647

Martino RL, Curran HA (1990) Sedimentology, ichnology and paleoenvironments of the upper cretaceous Wenonah and Mt. Laurel Formation, New Jersey. J Sediment Petrol 60(1):125–144

Milliman J (1972) Atlantic Continental shelf and slope of the United States-petrology of the sand fraction of sediments, northern New Jersey to southern Florida. US Geol Surv 529-J:Jl–J40

Montaggioni LF (1981) Les associations de Foraminife'res dans les se'diments re'cifaux de l'Archipel des Mascareignes (Oce'an Indien). Annales de l'Institut Oceanographique, vol 57, p 41–62. (Monaco)

Nelson CS (1988) An introductory perspective on non-tropical shelf carbonates. Sed Geol 60:3–12

Pettijohn FJ (1975) Sedimentary rocks, Harper and Row, New York, p 628

Scholle P, Ulmer-Scholle D (2003) A color guide to the petrography of carbonate rocks: grains, textures, porosity, diagenesis. American Association of Petroleum Geologists Bulletin Memoir 77, p 474

Simone L, Carannante G (1985) Evolution of a carbonate open shelf up to its drowning: the case of the Southern Apennines. Rend Accad Sci Fis Mat Napoli Ser IV 53:1–43

Thayer PA, Smits AD, Parker WH, Conner KR, Harris MK, Amidon MB (1993) Petrographic analysis of mixed carbonate-clastic hydrostratigraphic units in the general separations area (GSA), Savannah River Site (SRS), Aiken, South Carolina (U). U.S. : Department of Energy Report WSRC-RP-94-54. Westinghouse Savannah River Company, Aiken, p 83

Wilson JL (1975) Carbonate facies in geologic history. Springer, Berlin, p 471

Chapter 5
Sequence Stratigraphy

Abstract A sequence stratigraphic interpretation is presented for the lower Miocene Moghra Formation, northwestern Egypt, based upon 18 measured sections and the mapping of physical surfaces in superb desert outcrop over a lateral distance of more than 30 km. The Moghra Formation, which is exposed in a series of south-facing escarpments, consists of ~200 m of estuarine and marginal marine sandstone, siltstone, shale, and minor limestone, arranged into a series of trans-gressive-regressive cycles bounded by regional erosion surfaces with <15 m of local erosional relief. These surfaces are partially armored by thin ferruginous pebble conglomerate containing petrified wood and bone fragments. The cycles, reaching thicknesses of up to 45 m each, are dominated by transgressive deposits. The lower part of each cycle consists of unconsolidated cross-stratified sand and sandstone with locally abundant vertebrate fossils, petrified logs, *Thalassinoides* and *Ophiomorpha* burrows. These deposits, which are interpreted as a backstepping estuarine channel complex, are capped by pervasively bioturbated *Ophiomorpha*-bearing sandstone beds with marine fauna. In some cases, the marine beds pass upwards into a thin interval of regressive shale and siltstone, but such deposits are commonly truncated by the next master erosion surface (sequence boundary). Taken together, the Moghra sequences exhibit an overall transgressive trend, culminating in the open marine Marmarica Limestone. Provisional ages, based on a combination of biostratigraphy and strontium isotope stratigraphy, range from 21–17 Ma. The deep-marine oxygen isotope record for this interval is highly cyclic, with a subtle shift (to less positive $\delta^{18}O$ values) that is consistent a small sea-level rise. However, regional subsidence is required to account for most of the observed stratigraphic thickness.

5.1 Introduction: Background and Rationale

Sequence stratigraphy is considered by many as one of the latest conceptual revolutions in the broad field of sedimentary geology (Miall 1995), revamping the methodology and providing deeper process understanding of stratigraphic analysis. Sequence stratigraphy applications cover a tremendous range, from deciphering

S. M. Hassan, *Sequence Stratigraphy of the Lower Miocene Moghra Formation*   109
in the Qattara Depression, North Western Desert, Egypt, SpringerBriefs in Earth Sciences,
DOI: 10.1007/978-3-319-00330-6_5, © The Author(s) 2013

the Earth's geological record in a more dynamic way than through conventional stratigraphy (because it includes an understanding of the controls governing the key sedimentary processes) to using this new methodology for improving the success of petroleum exploration and production. Multiple data sets are integrated for this purpose, and insights from a variety of disciplines are required (Fig. 5.1). Sequence stratigraphy is uniquely focused on analyzing changes in facies and the geometric character of stratal units, identification of key surfaces across which there are shifts in stratal stacking patterns and then interpreting the relative roles of sediment supply, accommodation and autogenic processes for the development of the studied stratigraphic interval. Stratal stacking patterns are the signals or response to the interplay of autogenesis with changes in rates of sediment flux and base level, and are commonly progradation, retrogradation, aggradation and down cutting. A recent breakthrough with consensus among a surprisingly large group of sequence stratigraphers (Catuneanu et al. 2009) brought agreement that there are only 3 fundamental stratal stacking patterns, namely 'transgressive', 'normal regressive' and 'forced regressive'. During this final debate (Helland-Hansen 2009) some would still prefer that these three terms were more objective and less interpretive' and 'regression with rising trajectory' regression with falling trajectory and transgression were suggested for the three terms. After decades of disagreement and terminological debate, finally putting the emphasis on descriptive stacking pattern and disassociating the systems tracts from sea-level behavior is truly a great step forward. Facies and facies successions are generic from an environmental perspective (i.e., they can be identified in different depositional settings), and may be included in systems tracts of several age-equivalent depositional systems. Sequence stratigraphy can be an effective tool for correlation on both local and regional scales. The method is now commonly utilized as the modern approach to integrated stratigraphic analysis, combining insights from all other types of stratigraphic as well as several non-stratigraphy disciplines (Fig. 5.1) (Catuneanu et al. 2009).

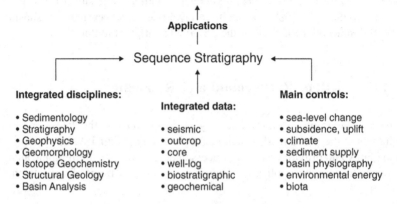

Fig. 5.1 Sequence stratigraphy in the context of interdisciplinary research (Catuneanu et al. 2009)

Over the past 30 yr, sequence stratigraphy has provided an important approach for evaluating the role of global sea level (eustasy), tectonic subsidence and uplift, and sediment supply processes on the deposition and outward growth of continental margin strata (e.g., Posamentier et al. 1988; Vail et al. 1977). Sequences are genetically related packages of sediment separated by unconformities or their correlative conformities (Mitchum Jr et al. 1977) and comprise the fundamental building blocks of the stratigraphic record (e.g., Christie-Blick 1991). Vail et al. (1977) and Haq et al. (1987) had suggested, at an early stage, that global sea-level (eustatic) change was the dominant process controlling sequences, though this soon brought protests that tectonic changes in base level also create sequence boundaries (e.g., Christie-Blick and Driscoll 1995). Eustasy and tectonics (including thermal subsidence, loading, flexure, and compaction) control accommodation, the space available for sediment to accumulate. Sediment supply controls how that space is filled. The interplay of accommodation and sediment supply control the formation of stratal surfaces, stratal geometries, and facies distributions as demonstrated by forward modeling (Reynolds et al. 1991; Browning et al. 2006).

Predictable, recurring sequences bracketed by unconformities comprise the building blocks of the stratigraphic record (Miller et al. 2004). At the very earliest stages of the debate a depositional sequence defined as a "stratigraphic unit composed of a relatively conformable succession of genetically related strata and bounded at its top and base by unconformities or their correlative conformities" (Mitchum Jr et al. 1977); by implication, "genetically related" referred to global changes in sea level (Vail et al. 1977). Christie-Blick and Driscoll (1995) clarified the genetic connotation by recognizing sequence boundaries as unconformities associated with lowering of base level, including eustatic and tectonic mechanisms. Sequences have been recognized in diverse stratigraphic environments (e.g., ranging from siliciclastic to carbonate setting; see examples in De Graciansky et al. 1998; Wilgus et al. 1988) from the Proterozoic (e.g., Christie-Blick et al. 1988) to the Holocene (e.g., Locker et al. 1996) and have been related to global sea-level (eustatic) variations (Haq et al. 1987; Posamentier and Vail 1988; Vail et al. 1977). However, differences in accommodation (the sum of subsidence and eustatic change) and sediment supply control sequence architecture, including the nature and significance of surfaces bounding and within sequences (e.g., flooding surfaces, transgressive surfaces). In addition, accommodation and sediment supply control the stacking patterns of facies successions within sequences (e.g., systems of Posamentier et al. 1988; Vail et al. 1977) and the general three dimensional arrangements of sequences (Reynolds et al. 1991).

One final recent discovery brings an interesting twist to this saga of understanding sequences. It turns out, after physical and numerical modeling studies (Muto and Steel 1997; Muto et al. 2007), that stratigraphic sequences at high-frequency level are not controlled merely by accommodation and supply, but also by unpredictable stacking-pattern changes even with constant forcing of supply and accommodation, i.e., by so-called autogenic responses. This has led to a proposal to try to explain stacking-pattern changes as far as possible with autogenic responses before resorting to unsteadiness in the forcing of accommodation and supply as an explanation for stacking changes. This method of analysis is known as Auto-stratigraphy (Muto et al. 2007), though it has been little used so far.

In the present study we use Sequence Stratigraphy as the study of sediments and sedimentary rocks in terms of repetitively arranged facies successions and associated stratal geometry (Catuneanu et al. 2009; Christie-Blick 1991; Vail 1987; Van Wagoner et al. 1988, 1990). We chose this approach not because we advocate or prefer a particular stratigraphic methodology (for a review see Catuneanu et al. 2009), but because it works well in our data set.

5.2 Sequence Architecture of Moghra Formation

Previous workers have not documented any sequence stratigraphy in the study area. By using outcrops and carefully measured stratigraphic sections we recognized at least eight to ten transgressive—regressive high-frequency cycles in Moghra Formation stratigraphy, identified them as sequences because of widespread erosion surfaces and abrupt vertical shifts of facies, and related them to accommodation change, presumably relative sea-level change as the strata are of marine and brackish water origin. We integrated newly-determined Sr-isotope data with biostratigraphy to provide improved age estimates for these sequences.

We provide: facies, facies associations, biostratigraphic, lithostratigraphic, and Sr- isotope data for identifying depositional sequences, documenting paleoenvironmental changes within sequences (e.g., facies tracts), and correlating sequences. We use this entire dynamic stratigraphic framework to evaluate the controlling mechanisms for the Early Miocene sequences in Moghra Formation, and if possible to extract a global sea level record for the area.

Despite great focus and considerable successes in understanding the vertebrates of the Moghra Formation, there is still a noticeable gap in our knowledge about the total dynamic stratigraphy of this area.

Applying depositional sequence stratigraphic concepts to the interpretation of siliciclastic depositonal systems has become increasingly important in petroleum geology. After a succession of break-throughs during the 1970 and 1980s, sequence stratigraphic concepts now have entered a phase of intense application and maturing. Workers have been using and integrating outcrops, cores, electric logs, and multifold seismic data. Clearly, sequence stratigraphic concepts embody not a rigid model or template, but rather a new way of looking at geology.

The objectives of this study are to (1) document the low- (third order, several my) and high-frequency (fourth-order, several 100 ky) stratigraphic sequences of the Mophra Formation, i.e., the dynamic stratigraphic framework of the Early Miocene succession within the study area and (2) elaboration of an Early Miocene sequence stratigraphic model in the study area.

Most of the studied exposures are good, but excavated trenches were dug in some sections of the outcrop. Our sections are closely spaced (less than 500 m to max 5 km), and we can walk many beds and surfaces along exposures (more than

Table 5.1 Summary of the main stratigraphic units and the sequence boundaries of the study area

Previous unit No's	Sequence No's	Basal sequence boundary
XVII	S11	SB11?
XVI	S10	SB10
XV	S9	SB9
XIV		
XIII	S8	SB8
XII		
XI	S7	SB7
X		
IX	S6	SB6
VIII		
VII	S5	SB5
VI		
V	S4	SB4?
IV	S3	SB3
III	S2	SB2
II		
I	S1	SB1 restricted in eastern part and unexposed rest of the sections.

35 km), facilitating correlation. The analysis of eighteen sedimentary sections located approximately along the depositional strike reveals complex stacking and juxtapositions of facies associations and stratal architecture for the seventeen units described in the previous chapter.

The following table gives an extended summary for the main stratigraphic units, sequences and sequence boundaries of the study area (Table 5.1).

Note that sequence boundaries define the base of the overlying sequence, and not necessarily the top of an underlying sequence due to irregular erosion on any SB.

5.3 Sequence Boundaries and Maximum Flooding Intervals

The conventional interpretation of **sequence boundaries** is that they are due to a relative fall of sea level and that they should develop more or less instantaneously. Christie-Blick and Driscoll (1995) argue that in many cases such boundaries actually form gradually over a finite interval of geologic time. Despite this, most researchers now accept that unconformity-bounded depositional sequences are approximate (albeit not exact) time units and represent the fundamental building blocks of sedimentary successions. These sequences are typically characterized by stratal onlap (transgression) at their base and by stratal offlap or downlap (regression) in their upper part, and they tend to be markedly asymmetrical, with onlap accounting for a larger part of any cycle of sedimentation compared to the downlapping portion. Offlap, or a falling trajectory near the upper boundary is caused by erosion and sediment by-pass and denotes the overlying sequence boundary, but its significance is greater than merely erosional truncation, as it also implies that sequence boundaries develop over an interval of time, i.e., during this period the downlapping clinoform migrates seawards and may be bypassing sediment downdip.

The interval or zone between the underlying onlapping and overlying downlapping units is referred to as the **maximum flooding interval.** The rock unit thus included between successive sequence boundaries, and including the maximum flooding interval, is known as the stratigraphic or depositional sequence. Depositional sequences always have a cyclic arrangement (transgressive–regressive) of facies, but it should be noted that these cycles cannot always easily be recognized downdip in deepwater areas. Here in down-dip, deepwater locations the sequence boundaries are both overlain and underlain by progradational deposits, and hence form during times of regression of the shoreline. By convention there are no 'transgressive' deposits in deepwater strata. Transgressive strata are confined to the shallow-water 'shelf'. At the landward end of depositional sequences transgressive units become thicker and dominate over regressive units because regression is recorded mainly by erosion and sediment bypass to the shoreline areas (Siggerud and Steel 1999). In the present study we are in an entirely shallow-water shelf setting, and transgressive deposits are of great and dominating importance, suggesting that we are sited on the landward reaches of a marine shelf.

Sequence boundaries (stratigraphic discontinuities, unconformities) are expressed by breaks in facies or biotic successions and by geometric patterns of stratal onlap and offlap that are in places directly observable in outcrop (e.g., Bosellini 1984, 1988; Christie-Blick 1991; García-Mondéjar and Fernández-Mendiola 1989, 1991; Mutti et al. 1985; Sarg 1988, 1989).

5.4 Sequence Boundaries and Maximum Flooding Intervals in Moghra Formation

In the Moghra Formation study we recognize sequence boundaries on the basis of physical features. Criteria for recognizing sequence bounding unconformities include: (1) irregular erosion surfaces, with up to 15 m relief or inferred relief; (2) evidence of reworking, including rip-up clasts found right above the contact; (3) heavy bioturbation, including burrows filled with overlying material as much as a metre below the contact; (4) a major lithofacies shifts typically from somewhat deeper to somewhat shallower water across the contact; (5) vertebrate lags above the contact; In general, most sharp contacts proved to be either sequence boundaries or maximum flooding surfaces (MFS). MFS may be differentiated from

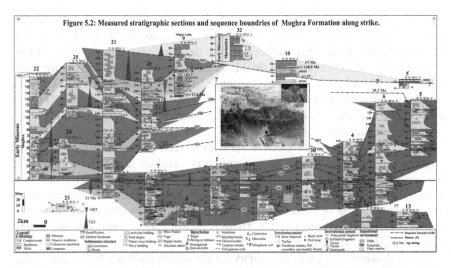

Fig. 5.2 Measured stratigraphic sections and sequence boundries of Moghra Formation along strike

sequence boundaries by its muddy or open marine character and because it represents the level of maximum water depth or deepest water facies compared to the overlying and underlying facies. Though MFSs are heavily burrowed they generally lack rip-up clasts and are associated with the uppermost part of distinct retrogradational lithofacies successions. Not all potential sequence boundaries display all of the criteria listed above, though the minimal evidence for a sequence boundary requires a sharp lithologic contact, a basinward facies shift, and evidence of erosion (rip-up clasts and lags) (Browning et al. 2006). We recognize eleven sequence boundaries (Fig. 5.2) within the Early Miocene Moghra Formation and these are supported by walk-out lateral field correlations, indicating that they can be correlated sub-regionally. These eleven sequence boundaries are regional erosion surfaces with <15 m of local erosional relief. These surfaces are partially armored by thin ferruginous pebble conglomerate containing petrified wood and bone fragments. These eleven main erosional surfaces within the Moghra Formation probably did not form by autocyclic processes. Their considerable incised relief, up to 15 m, their association with conglomerate basal lags, and the great total thickness of the sandstone body suggest that these erosion surfaces originated as allocyclic responses to unsteady forcing of supply or sea level, creating base-level fall events (SBs) followed by high-frequency transgressive–regressive movements of the shoreline while sea level was rising.

The maximum flooding interval is represented by heavily burrowed glauconitic beds and more diverse assemblages of fossils (including abundance of peleypods, echinodermata and oysters). Maximum flooding intervals in parts of the stratigraphic succession are more problematical, particularly where they are within successions lacking regressive facies tracts, and where the overlying SB may have eroded right down into the transgressive tract. Vertically above the SB and within the transgressive tract, there are several important transgressive surfaces that occur, especially in transgressive estuarine environments like this. We tentatively place Rt (tidal ravinement surface) within the proximal transgressive estuary deposits (Fig. 3.25), dividing the tidal-influenced fluvial channels below from the tidal channels and tidal bars above, in most of the sequences. Because LST are generally absent (though the occasional interval of fluvial deposits below Rt could possibly be classed as lowstand deposits), transgressive surfaces are sometimes merged with the sequence boundary. Rw (wave ravinment surface, the second, more seaward type of transgressive surface) is recorded at the top of the tidal channel and tidal bars facies association and is a surface immediately below the lag accumulation of oysters or open marine fossils/deposits (Figs. 3.24, 3.25). This wave ravinement surface is where there is usually some wave action in front of the estuary mouth, causing the accumulation of fragments of marine fossils and small pebbles (Siggerud and Steel 1999). There is rapid fining of grain size above Rw, into the muddier MFS interval.

The following table gives a summary for the eleven sequence boundaries of the study area (Table 5.2).

Table 5.2 Summary for the sequence boundaries of the study area

Sequence boundary No.	The most characteristics
SB1	This sequence boundary is underlain by covered deposits and overlain by tide-dominated inner estuary (tide-influenced fluvial channel) deposits
SB3, SB5, SB6 & SB8	These sequence boundaries mark an abrupt basin ward shift in sedimentary facies. The shift is from estuary and open shelf deposits (sandstones with moderately to intensively bioturbated branched network *Ophiomorpha)* below up to tide dominated estuary (tide-influenced fluvial channel at the base, grading upwards into, tidal channels and tidal bars) above
SB7	This sequence boundary marks an abrupt basinward shift in sedimentary facies from tide-dominated estuary margin deposits interpreted as tidal-flat deposits below, up to tide-dominated estuary (tide-influenced fluvial channel) above
SB2, SB9 & SB 10	These sequence boundaries mark a shift in sedimentary facies from regressive tide-dominated delta deposits (both prodelta and delta front) below up to tide dominated estuary (tide-influenced fluvial channel at the base) above
SB4 & SB11 (act also as MFS)	These sequence boundaries mark an abrupt shift in sedimentary facies from tide dominated estuary tidal flat facies association (heterolithic beds and inclined heterolithic stratification) below up to open shelf deposits above

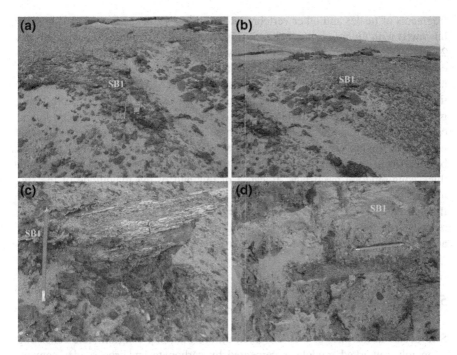

Fig. 5.3 (**a & b**) The first sequence boundary SB1, based the first depositional sequence S1 lined by lag deposit, bone fragments and other vertebrate fossils. (**c**) Fragmented silicified wood on the erosional surface. (**d**) Visible trace fossils are *Ophiomorpha* below the sequence boundary

5.4.1 Moghra Formation SB1

This sequence boundary does not have a widespread extent across the study area. In the western areas it is not at all preserved. This surface is poorly preserved in two sections further east (Sections. 26 & 13) and not exposed at all in other sections. SB 1 is lined by disordered lag deposits, silicified wood but lacks any bone fragments.

5.4.2 Sb2

SB 2 is marked at the base by 30 cm thick highly ferruginous, intraformational conglomerate that is disordered and matrix (coarse-grained sandstone) supported. The clasts are mainly highly ferruginated mudstone but there are also abundant quartz pebbles, granules and vertebrate bone fragment (Fig. 5.3a, b & c). This lag deposit lying on the erosional surface contains fragmented silicified wood, bone fragments of turtles and other vertebrate fossils. The sequence boundary is heavily burrowed with *Ophiomorpha* below it (Fig. 5.3d). SB 2 separates sequences 1 and 2 and marks an abrupt shift in sedimentary facies. The shift is from regressive tide-dominated delta deposits (both prodelta and delta front) below up to tide dominated estuary (tide-influenced fluvial channel at the base, grading upwards into, tidal channels and tidal bars) above (Fig. 5.2). In few section it separate transgressive estuary below from tide dominated estuary above (tide-influenced fluvial channel). This erosional surface at the base of the sandstone body of the tide-influenced fluvial channel was probably not formed by autocyclic processes because of its considerable incised relief (in excess of 10 m) and association with thick conglomerate basal lags. Because of this and of the magnitude of the significant basin ward shift of facies (from muddier delta deposits up to sand-rich fluvial-tidal deposits) it is likely that this erosion surface originated as allocyclic responses to unsteadiness in the sea-level or supply forcing, and was followed by accumulation of a high-frequency transgressive–regressive cycle.

5.4.3 Sb3

SB3 occurs at the base of a 10 cm-thick intraformational conglomerate consisting of mainly reworked mudclasts cemented by carbonate and iron oxides cement. This sequence boundary separates sequences 2 and 3 (Fig. 5.4 a & b) and also marks an abrupt basin ward shift in sedimentary facies. The shift is from estuary and open shelf deposits (sandstones with moderately to intensively bioturbated branched network *Ophiomorpha)* below up to tide dominated estuary (tide-influenced fluvial channel at the base, grading upwards into, tidal channels and tidal bars) above. This sequence boundary represents a regionally extensive surface from east to the west direction of these depositional strike sections.

5.4.4 Sb4

This surface proved to be either sequence boundary (SB4) or maximum flooding surface (MFS) (similar case see Browning et al. 2006). We tentatively place SB at this level. This surface could be marked the MFS overlie the sequence boundary? However, this sequence boundary seems an ambiguous surface; it is clearly seen in section 30 and marked by lag deposits and mudclasts (Fig. 5.5a). On the other hand, it coincides with the maximum flooding surface in the rest of the sections. In addition, this sequence boundary separates sequence 3 and 4 and marks an abrupt shift in sedimentary facies. The shift is from tide dominated estuary, marginal marine, tidal flat facies association (Fig. 5.5a) (heterolithic bed and inclined

Fig. 5.4 (a & b) The
sequence boundary SB2,
based the depositional
sequence S3 and overlain by
inclined heterolithic facies
and underlain by tidal sand
bar

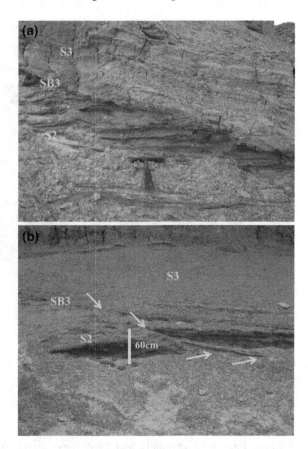

heterolithic stratification and mudstone bed) below up to open shelf deposits (this
Bioturbated—fossiliferous sandstone facies association FA5 subdivided into
1-fossiliferous calcareous sandstone, highly bioturbated with *Ophiomorpha* and
Thalassinoid, and 2- calcareous, mottled, homogenized, highly bioturbated
Ophiomorpha sandstone) above. The extension of this surface in the entire study
area is considered as puzzle and need further future work.

5.4.5 *Sb5*

SB 5 is marked at the base by 10-20 cm thick lag conglomerate (Fig. 5.5b & c).
This sequence boundary separates sequences 4 and 5 and marks an abrupt shift in
sedimentary facies. The shift is from open shelf deposits (fossiliferous calcareous
sandstone, highly bioturbated with *Ophiomorpha* and *Thalassinoid*) below up to
tide dominated estuary (tide-influenced fluvial channel at the base, grading

Fig. 5.5 (a) Sequence boundary SB4, based depositional sequence S4 overlain by open shelf deposits and underlain by tidal flat facies Association. (**b** & **c**) Close view to the mudclasts above sequence boundary SB5. (**d**) Sequence boundary SB5 at the base of depositional sequence 5 overlain tidal channel tidal bars and SB6 at the *top*

upwards into, tidal channels and tidal bars) above (Fig. 5.5d). This sequence boundary has a widespread extent across the eastern and western part of the study area with great variability in erosion below it.

5.4.6 Sb6

SB 6 is marked by highly ferruginous lag gravels and silicified wood fragment, coprolites and bone fragments (Figs. 5.5d, 5.6a & b). This sequence boundary separates sequences 5 and 6 and marks an abrupt basin ward shift in sedimentary facies. This shift is from open shelf deposits (fossiliferous calcareous sandstone, highly bioturbated with *Ophiomorpha* (Fig. 5.6c) and *Thalassinoid* (Fig. 5.6d) below up to tide dominated estuary (tide-influenced fluvial channel at the base, grading upwards into tidal channels and tidal bars) above. This sequence boundary has a widespread extent across the eastern and western part of the study area but less extension than SB5. In addition, it has not great variability in erosion below it like SB5.

Fig. 5.6 (**a** & **b**) Sequence boundary SB6, based the depositional sequence and lined by highly ferruginated lag gravels and silicified wood fragment, coprolites, bone fragments. (**c**) Close view of the *Ophiomorpha* trace fossils of fossiliferous calcareous sandstone, highly bioturbated with *Ophiomorpha* above sequence boundary SB6. (**d**) Close view of *Thalassinoid* trace fossil above SB6

5.4.7 Sb7

SB7 is marked by highly ferruginous lag gravels and silicified wood fragment, coprolites, bone fragments of the turtles, crocodile and other mammalians (Fig. 5.7a, b & c). This sequence boundary separates sequences 6 and 7 and marks an abrupt basin ward shift in sedimentary facies. This shift is from tide-dominated estuary margin deposits with preserved heterolithic intervals interpreted as tidal-flat deposits (Fig. 5.7c) and mudflat deposits below up to tide dominated estuary (tide-influenced fluvial channel at the base, grading upwards into, tidal channels and tidal bars) above. This sequence boundary has a widespread extent across the eastern and western part of the study area with great variability in extent and magnitude of erosion below it.

5.4.8 Sb8

SB8 is marked by small petrified trunks and fragments of silicified wood and vertebrates bone fragments. This sequence boundary separates sequences 7 and 8 (Fig. 5.7d) and marks an abrupt shift from open shelf deposits (fossiliferous

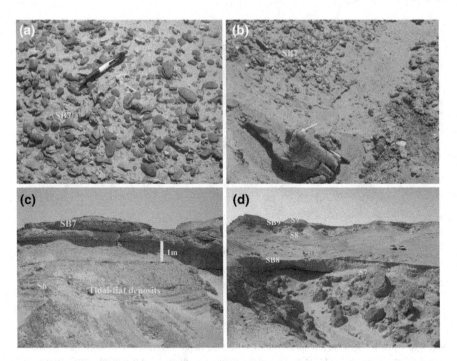

Fig. 5.7 (**a**) Close view of crocodile bone fragment and highly ferruginated lag gravel above SB7. (**b**) Close view of giant of mammalians bone fragment (*yellow arrow*) that found over SB7. (**c**) Sequence boundary SB7 underlain by heterolithic bed of tidal flat deposits of sequence 6. (**d**) The sequence boundary SB8 based the depositional sequence S8 and overlained by SB9

calcareous sandstone, highly bioturbated with *Ophiomorpha* and *Thalassinoid*) below up to tide dominated estuary (tide-influenced fluvial channel at the base, grading upwards into, tidal channels and tidal bars) above. This sequence boundary has a widespread extent especially across much of the western study area with great variability in erosion. In the eastern areas it is less widespread or has limited extension (only section5).

5.4.9 Sb9

SB9 is marked by a concentration of intraformational conglomerate, wood, bone and coprolites. This sequence boundary separates sequences 8 and 9 and is inferred by a sharp lithofacies shift. The shift is from regressive tide-dominated delta deposits (both prodelta and delta front) below up to tide dominated estuary (tide-influenced fluvial channel at the base, grading upwards into, tidal channels and tidal bars) above. This sequence boundary has a widespread extent especially across much of the western study area with less erosion (present in 4 sections, 20, 21, 22 and 25). In the eastern areas it is not preserved at all.

5.4.10 Sb10

SB10 is marked by an erosional surface with intraformational conglomerate, silicified wood trunks, coprolites and bone fragments. This sequence boundary separates sequences 9 and 10 and is inferred from a sharp lithofacies shift. The shift is from regressive tide-dominated delta deposits (both prodelta and delta front) below up to tide dominated estuary (tide-influenced fluvial channel at the base, grading upwards into, tidal channels and tidal bars) above. This sequence boundary has a limited extent especially across much of the western study area. As with SB9 it is not preserved in the eastern area.

5.4.11 SB11 (between Marmrica and Moghra Formation)

SB 11 resembles surface SB3. It proved to be either sequence boundary (SB11 or maximum flooding surface (MFS) (similar case see Browning et al. 2006). So, we tentatively place an SB or MFS at this level. This surface is characterized by coarse grained, pebbles, small pelecypod and mud clasts (Fig. 5.8a). This sequence boundary marks an abrupt shift in sedimentary facies. The shift is from transgressive estuary (tide-influenced fluvial channel at the base, grading upwards into, tidal channels and tidal bars) below up to open shelf deposits (bioturbated-fossiliferous carbonate) above (Fig. 5.8b). This surface could be characterized by MFS overlying and amalgamated with the sequence boundary. Regressive facies tracts are not present in most of sections below this surface, but the transgressive interval is shifted sufficiently basin-ward from the preceding interval. In fact, this surface for first glance looks like it belongs to Marmarica Formation because the limestone bed above it shows characteristic features of Marmarica Formation that are known from the previous literature and mentioned early in the previous work of the introduction chapter, but the result from the age dating of samples of this bed gives evidence of Burdigalian age. This means that the Marmarica Formation dates to the Lower Miocene (Burdigalian age).

5.5 Description of Depositional Sequences and Component Cycles

Work on the Moghra Formation has documented a series of high-frequency (fourth-order) sequences 20–60 m thick, each bounded below by an erosive sequence boundary as discussed above. All sequences are characterized by a thick (as much as 54 m, section 6) cross-stratified sandstone-dominated succession in their lower parts. This thick basal sandstone is then gradually overlain by the incoming of *Ophiomorpha* beds and becomes gradually finer grained strata, before

Fig. 5.8 (a) Close view of pebbles (*yellow arrow*) and mud clasts above SB11. (a) The sequence boundary SB11 based the depositional sequence S11 overlain by open marine deposits

being capped by thin calcareous marine beds and mudstones in places. In some sections there is some grain size reversal with an irregular upwards coarsening and increasing sand beds in the capping succession before the next sequence boundary. Age control for the early Miocene sequences is provided by an integration of Sr-isotope stratigraphy and conventional biostratigraphy. Sr-isotope based age estimates were obtained from mollusk, echinoid and pectin carbonate shells. The Moghra Formation is thus asymmetric-cyclic, characterized by a repetitive series of stratigraphic sequences, each composed of an erosively bounded, thick transgressive facies tract and a thin or absent regressive facies tract. The sand-rich transgressive tract is typically composed of a basal fluvial unit with some tidal influence (sometimes this basal sand is absent), an erosive tidal ravinement (Rt) surface at the incoming of cross-stratified sandstones formed in tide-dominated channels, then a wave ravinement (Rw) at the level of appearance of open-marine fossil lag accumulations with some overlying upward-fining sandstones and eventually by a maximum flooding surface within muddier facies. This transgressive tract is interpreted in terms of a backstepping of facies associations: tidal-fluvial deposits, then inner estuary and eventually outer estuary channel-bar deposits (mixed siliciclastic and carbonate facies). Regressive systems tracts are bounded beneath by an MFS (and above by a SB) and characterized by fore-stepping regressive tide-dominated delta (prodelta and front delta) or intertidal deposits. The regressive (highstand) tracts are partially truncated by the next sequence boundary. Together, these sequences constitute the component transgressive and regressive (highstand) tracts of a larger (third-order Miocene) composite sequence. No lowstand systems tracts have been recognized within the succession, though some occasional basal fluvial deposits (and certainly we would expect these to expand in the unexposed upstream reaches of this system) may represent lowstand periods.

5.5.1 Sequence 1 (4–30 m thick; Fig. 5.9a)

Sequence 1 reaches 30 m thick (sections 7 & 30) and is erosionally reduced to 4 m in places (section 6, Figs. 2.1 & 5.2). The sequence is based by SB1 which it is preserved as traces only in the eastern part from the study area, sections 13 an 26 (Figs. 2.1 & 5.2). This sequence is characterized by preservation of transgressive tract and regressive tract. The transgressive tract represented by 20–30 m of tide-influenced fluvial channel in the eastern studied area (sections 26 & 13, Fig. 2.1). In addition, it is represented by 10–15 m by marginal estuary tidal flat FA4 (sections, 8, 1, 2 and way point 307, Figs. 2.1 & 5.9a) and is capped by the irregular erosive boundary of overlying sequence 2. The thick regressive tract of Sequence 1 consists of irregular but crudely upward-coarsening 30 m-thick (section 7, Fig. 2.1), tide-dominated delta deposits. There are three facies in this deltaic unit: 1- coarse-grained lensoidal heterolithic at the top representing upper delta-front deposits, 2- thin laminated sand-shale intercalation in the middle as main delta-front and (3) bioturbated heterolithic sandstones and shales as prodelta deposits. The occurrence of a regressive deltaic systems tract in any sequence has the pre-condition that the backstepping estuary was able to entirely infill the eroded valley area during transgression (Dalrymple 1992). The maximum flooding interval of sequence 1 (MFI) is very difficult to determine. Sequence 1 is dated with Sr-isotopic data to 21 Ma in the uppermost part of the sequence.

5.5.2 Sequence 2 (12–40 m thick; Figs. 2.1 and 5.2)

Sequence 2 varies from 40 m thick (section 1) to 12 m (section 21) because of the erosive character of overlying SB 3 (Figs. 2.1, 5.2). Sequence 2 consists of a transgressive facies tract, normally deepening upwards, dominated by Facies Associations 1 and 5 (inner estuary deposits, tidal-fluvial deposits and open shelf, Bioturbated fossiliferous sandstone FA5). This transgressive tract consists of 16 m thick, erosionally based, fine-to coarse-grained, fining-upward sandstone bodies with abundant clay clasts, quartz pebbles, lithic fragments, plant debris, and occasional tree chunks and vertebrates bones fragments that represent the fluvial-tidal channels of the inner-most estuary zone (Facies Association 1). This facies association graded upward to facies associations 2 and 3 that represent by middle to outer estuarine tidal sand bars. This transgressive tract grades upwards to open shelf deposits that include bioturbated-fossiliferous sandstones with moderately to intensively bioturbated branched networks of *Ophiomorpha* that represent the maximum flooding interval or close to it. There are no regressive deposits preserved in sequence 2 because the overlying sequence boundary 3 cuts down severely to truncate the entire upper part of this sequence, as it does in many of the other sequences in the study succession. In this case the overlying regressive tract may simply have not been developed probably where the incised estuaries were not able to be completely filled during the transgressive phase. Kidwell (1997) has referred to such incomplete T-R sequences as 'shaved' sequences; others simply refer to them as truncated sequences. The maximum flooding interval 2 (MFI 2) is represented by 4 m to 2–3 m (S1& S4) sandstone with moderately to intensively bioturbated branch network *Ophiomorpha* (Fig. 5.9a & b). In other sections it is very hard to know (For example, S2, S5 and S21). Sequence 2 is dated with Sr-isotopic data to 19.6 Ma, from samples taken in the middle of the sequence.

5.5.3 Sequence 3 (2–20 m thick; Figs. 2.1 and 5.2)

Sequence 3 is about 20 m thick (section Aux) and reduces to 2 m in section 21. Sequence 3 consists of both transgressive facies tract and very thin regressive facies tract (Fig. 5.10 a & b). The transgressive tract is variable from section to section and is represented by 6–12 m (sections 6 & 7) of tide dominated estuary deposits that include axial facies associations: tidally influenced fluvial channels and overlying outer estuary tidal channel-bar systems. These outer estuary bars contain three facies: 1- tabular and trough cross stratified sandstone (2-D and 3-D dunes), 2- lensoidal calcareous cross-stratified sandstones, and 3- sandstones with *Ophiomorpha* burrows. In addition in some sections this transgressive facies tract is represented by 8 m (section 1) of tidal flat deposits (not axial estuary as above but muddier margin estuary deposits). This marginal facies association can be subdivided into heterolithic beds (8 m, section 1) and inclined heterolithic

Fig. 5.9 (a) Panoramic view of Moghra formation section illustrating sequence 1 of marginal estuary heterolithic tidal flat deposits of transgressive tract of this sequence. (b) Panoramic view of transgressive tract of sequence 3, heterolithic tidal-flat of S3 overlain by transgressive tract of tidal sand bar of S 4. (b & c) The maximum flooding interval 2 (MFI 2) of sequence 2 is represented by sandstone with moderately to intensively bioturbated branch network *Ophiomorpha*. Note the pebbles lined the burrow

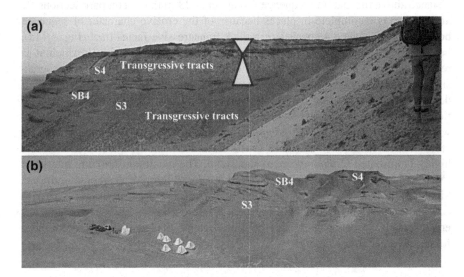

Fig. 5.10 (a & b) Panoramic view of Moghra formation section illustrating relationships between the heterolithic tidal-flat of S3 overlain by tidal sand bar of S 4 deposits and how they correspond to systems tracts. The transgressive facies tract is represented by 8 m (section 1) of tidal flat deposits (muddier margin estuary deposits). This marginal facies association can be subdivided into (a) heterolithic beds (8 m, section 1) and (b) inclined heterolithic stratification and mudstone beds

stratification and mudstone beds (18 m, section Aux). The regressive tract may simply have been poorly developed in one location (section 4) and not developed at all in other sections. The poorly preserved, very thin regressive tract of this sequence consists of upward-coarsening 4 m-thick (section 4), tide-dominated delta deposits that include coarse-grained lensoidal heterolithic sandstones at the top representing upper delta-front deposits. This sequence is severely eroded by SB 4, so that regressive facies tracts are lacking, either because regression was non-depositional or because regressive deposits were eroded during the final phase of regression and/or during the next transgression. The maximum flooding interval 3 (MFI 3) is not easily recognized in any section. Sequence 3 has a widespread extent (present in 7 sections from east to the west and cover around 20 km) especially across much of the eastern study area with great variability and thickness. In the western areas it is less widespread. This sequence lacks age control because the general lack of in-place fossils severely limits dating of this sequence.

5.5.4 Sequence 4 (2–28 m thick; Fig. 5.10)

Sequence 4 is somewhat problematic because it is bounded by SB4 below and by SB5 above. SB4 appears to be, at first glance, a maximum flooding surface that is used as datum for the correlation panel. However, SB4 is likely to be a short distance above this datum. Sequence 4 is about 2–28 m thick (compare sections 21 and 30).but is incomplete in places because of the erosive character of sequence boundary 5. Sequence 4 consists of a thick transgressive facies tract (Fig. 5.11a), but its regressive facies tract is not preserved except in section 1 which it is represented by 2 m coarsening upward deltaic facies (Fig. 5.11b). The transgressive facies tract consists of 28 m (section 30) of deposits that represent open shelf deposits. These consist of the bioturbated-fossiliferous sandstone facies association that includes facies: 1- Fossiliferous calcareous sandstone, highly bioturbated with *Ophiomorpha* and *Thalassinoides*. 2- Fossiliferous ferruginous large-scale cross-bedded sandstone with hard crust, calcareous, mottled, homogenized and highly bioturbated (*Ophiomporpha*) sandstone. The maximum flooding interval 4 (MFS 4) is represented by 6–7 m (section 30) fossiliferous calcareous sandstone, highly bioturbated with *Ophiomorpha* and *Thalassinoide*s. It is very hard to determine in other sections because there is more incision and severe sequence erosion. This sequence lacks age control because the general lack of in-place fossils severely limits dating.

5.5.5 Sequence 5 (8–48 m thick; Figs. 2.1 and 5.2)

Sequence 5 is about 48 m thick in section 6 and reduces to 8 m in Section 22, and is bounded below by the SB5 and above by SB6. This sequence consists of a very

Fig. 5.11 (a) Panoramic view is showing the transgressive facies tract that represent open shelf deposits. These consist of the bioturbated-fossiliferous sandstone facies association FA 4 of Sequence S4 and topped by Sequence boundary SB5. Note that the thickness of this facies increase toward the west and decrease towards the east direction (look for the variable thickness of the arrows). (b) Photographs of regressive tract of sequence 4 and represented by coarsening upward deltaic facies

thick transgressive facies tract and very thin regressive facies tract. In most of the area the regressive deposits are thin or lacking, as are any regressive deposits, and so this sequence too is the remnant of severe erosion. The transgressive tract in section 6 is represented by 48 m of estuarine deposits that include axial facies associations: tidally-influenced fluvial channel and outer estuary tidal channel and bar deposits. This latter association includes three facies: 1- tabular and trough cross stratified sandstone, 2- lensoidal calcareous cross-stratified sandstone (Fig. 5.12a) and 3- *Ophiomorpha* burrows. The top-most part of the facies tract represents open shelf deposits some 10–12 m thick (section 6). The open shelf deposits includes bioturbated—fossiliferous sandstone which it can be subdivided into facies as follow: 1- Highly bioturbated sandstone with honey comb. 2- Calcareous, mottled, homogenized, highly bioturbated with *Ophiomporpha* and *Thalassinoides*. The overlying regressive tract is preserved only in sections 22 and 24 towards the west of the study area and a thin remnant in section 5 towards the east of the study area. The regressive tract of Sequence 5 consists of upward-coarsening 20 m-thick (section 24), tide-dominated delta deposits. The basal part of these deposits is coarse grained lensoidal heterolithic. This regressive tract grades upward to a fining—upward tidal bar deposits. The top-most part of the regressive tract shows 8 m coarsening-upward, homogenized bioturbated sandstone that may represent prodelta deposits. The maximum flooding interval 5 (MFI 5) is

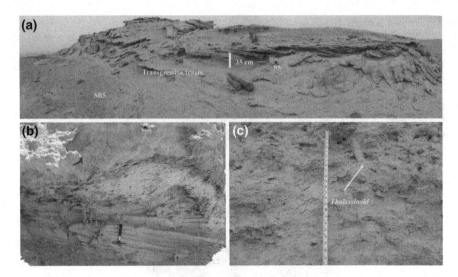

Fig. 5.12 (**a**) Panoramic view is showing Sequence S5 that consists of transgressive tracts which it represented by of tide dominated estuary deposits that include axial facies associations: lensoidal calcareous cross-stratified sandstone (*yellow arrow*). (**b**) Photographs has illustrated the maximum flooding interval 5 (MFI 5) is represented by fossiliferous calcareous sandstone, highly bioturbated with *Ophiomorpha*. (**c**) Close view of the fossiliferous sandstone and include the *Thalassinoid* trace fossils (*Yellow arrow*)

represented by 6 m (S6) highly bioturbated sandstone with honey comb and calcareous, mottled, homogenized, highly bioturbated (*Ophiomorpha*) sandstone and represented by 2.5 m (S21) fossiliferous calcareous sandstone, highly bioturbated with *Ophiomorpha* and *Thalassinoid* (Fig. 5.12b & c). It is very hard to determine in the other section because severely erosion. No significant delta- influenced facies are visible. This sequence lack age control because the general lack of in-place fossils for dating.

5.5.6 Sequence 6 (12–56 m thick; Figs. 2.1 and 5.2)

Sequence 6, about 56 m thick in section 6 and reduces to 10-12 m in sections 4 and 22, is bounded by SB6 below and by SB7 above. This sequence consists of a very thick transgressive tract and a thinner, coarsening-upward regressive tract. The transgressive tract is some 60 m thick in section 21, and is interpreted as an estuarine succession that includes axial facies associations: tide-influenced fluvial channel and outer estuary bar (tidal channel and tidal bars) deposits. The outer estuary deposits can be subdivided into three facies: 1- tabular and trough cross stratified sandstone, 2- lensoidal calcareous cross-stratified sandstone and 3- sandstone with *Ophiomorpha* burrows. The topmost part of this tract represents open shelf deposits. The thinner regressive facies tract is some 8 m thick in

Fig. 5.13 (a) Photomosaic showing slope forming sandstone of S5 overlained by heterolithis tidal-flat of S6 and terminated by fossileferous limestone open Marin bed of S7. (**b**) Panoramic view is showing heterolithic tidal-flat of S6. (**c**) Original photomosaic showing S7at the base that represented by heterolithic at the base and topped by slop forming cross-bedded sandstone of tidal channel and terminated by fossiliferous limestone

section 21 and consists of a coarsening upward, regressive tide-dominated delta (Fig. 5.13 a, b & c). This facies association is capped by coarse-grained lensoidal heterolithic deposits. The maximum flooding interval 6 (MFI6) is restricted in by 6 m in section 20 and interpreted a more diverse fossiliferous limestone with fossils assemblage associated with the MFS (including abundant peleypod, echinodermata and pecten). This fossiliferous limestone represents the deepest part in the sequence. In addition to intensive sandstones with moderately to intensively bioturbated branched network *Ophiomorpha*? It is very hard to determine this interval in other section could be the marine sand body and regressive tide dominated of prodelta and delta front are inter-fingered there so it is very hard to differentiate it or could be eroded in other sections. This sequence lack age control because the general lack of in-place fossils severely limits chronostratigraphic control of this sequence.

5.5.7 Sequence 7 (9–60 m thick; Figs. 2.1 and 5.2)

Sequence 7, about 60 m thick (section 21) and reduced to 9 m in section 20 (Figs. 2.1 & 5.2), is bounded below by the SB7 and above by SB8. This sequence

Fig. 5.14 (a & b) Photomosaic showing slope forming sandstone of S7 graded up into fossileferous limestone open marine bed that topped by slop forming cross-bedded sandstone of tidal channel of S8 that bounded by SB9 at the top and overlained by S9

consists of a very thick transgressive tract and very thin to absent regressive tract. The transgressive tract is represented by 60 m thick in section 21 (Fig. 5.14 a & b), and is interpreted as an estuary succession that includes axial facies associations: tidal- influenced fluvial channel and outer estuary bar (tidal channel and tidal bars). The top-most part of this facies tract represents open shelf deposits that include the bioturbated-fossiliferous sandstone facies association. This facies association can be subdivided into facies as follow: 1- Sandstones with moderately to intensively bioturbated branched network *Ophiomorpha*, 2- Fossiliferous ferruginous large scale cross-bedded sandstone with hard crust. This sequence is also severely eroded (i.e., partial truncation of the sequence, removing part or all of the regressive facies tracts and possibly part of the transgressive facies tract). So, this sequence represents 'shaved' sequence in the terminology of Kidwell (1997). However, there is a 6 m section 22 remnant of regressive faces tracts represented by heterolithic sandstone that could be marginal estuary, tidal flat facies association of estuary or could represent coarsening upward—heterolithic facies of regressive tide-dominated delta. The maximum flooding interval (MFI) is widespread through the western study area. The maximum flooding interval 7 (MFI 7) is represented by 12 m (section 20) fossiliferous limestone. The bioclastic fragments (e.g.oyster and echinodermata) are commonly distributed in this interval and going up to fossiliferous calcareous sandstone, highly bioturbated with *Ophiomorpha* and *Thalassinoid*. The lowest thickness of this interval is 2 m (section 20) sandy fossiliferous limestone. It is also represented by 12 m thick (section 25) moderately bioturbated fossiliferous limestone (oyster) and the trace-fossil assemblage is dominated by *Thalassinoides* and going up to sandstones with

moderately to intensively bioturbated branched network *Ophiomorpha* and fossiliferous limestone (?). Sequence 7 is dated with Sr-isotopic data to 18.2 Ma,
from samples in the most top part of the sequence.

5.5.8 Sequence 8 (12–56 m thick; Figs. 2.1 and 5.2)

Sequence 8 is about 52 and 56 m thick (sections 22 & 25) and reduces to 12 m in
section 20. This sequence is bounded by the SB8 below and by SB9 above and has
moderately thick transgressive and regressive facies tracts. The transgressive
facies tract is some 28 m thick in section 25, and is interpreted as an estuary
succession that includes axial facies association (tidal channel and tidal bars).
There are several internal erosional surfaces within the sandstone that probably
were formed by autocyclic processes. The topmost part of this facies tract represents open shelf deposits (Fig. 5.15 a, b & c). These open shelf deposits include the
bioturbated-fossiliferous sandstone facies association that can be subdivided into
facies as follow: 1- Sandstones with moderately to intensively bioturbated branched network *Ophiomorpha*, 2- fossiliferous calcareous sandstone, highly bioturbated with *Ophiomorpha* and *Thalassinoid*. The regressive facies tract is some

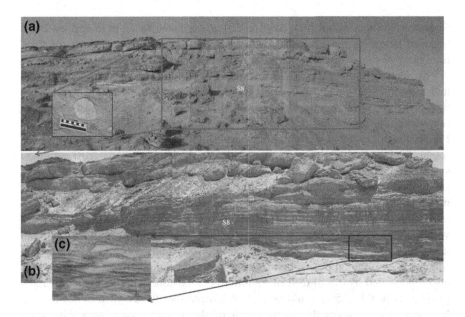

Fig. 5.15 (a) Panoramic view of Moghra formation section illustrating relationships between the
heterolithic tidal flat of S8 overlain by lag deposits and intraclasts of SB 9 that topped by tidal
channel S9. (b) Original photomosaic showing S7 at the base that represented by heterolithic at
the base and topped by slop forming cross-bedded sandstone of tidal channel and terminated by
fossiliferous limestone. (c) Close view for the trough cross-bedded sandstone (shovel for scale)

20 m thick in section 22 and consists of a coarsening upward, regressive tide-dominated delta, thin laminated sand- shale intercalation (prodelta and delta front deposits). This facies tract is reduced in thickness laterally into 8 m in section 21 and represented by facies coarse grained lensoidal heterolithic with tide dominated delta. The maximum flooding interval 8 (MFI 8) is represented by 13 m (Section 22) calcareous, mottled, homogenized, highly bioturbated (*Ophiomorpha*) sandstone and going up to intensively bioturbated fossiliferous sandstone. This interval is also represented by 8 m (section 25, Fig. 5.15 a & b) of sandy fossiliferous limestone and fossiliferous limestone bioclastic fragments (e.g. bivalves, gastropods, echinodermata). Intensity of bioturbation is commonly high and the trace-fossil assemblage is dominated by *Ophiomorpha* sandstone intervals. This interval is represented laterally by 8 m (section 21) of intensively bioturbated glauconitic sandstone. Furthermore, this interval is represented in the east by only 2 m thick (Section 5) karstified fossiliferous limestone. Sequence 8 has a widespread extent (present in 7 sections from east to the west) especially across much of the western study area with great variability and thickness. This sequence lacks age control because of the general lack of in-place fossils for age dating.

5.5.9 Sequence 9 (21–50 m thick; Figs. 2.1 and 5.2)

Sequence 9, about 50 m thick (section 20) and reduced to 21 m in section 21, is bounded below by the SB9 and above by SB10. This sequence consists of thick transgressive facies tract (28 m, section 22) and very thick regressive facies tract (44 m, section 9). The transgressive tract is interpreted as an estuary succession that includes axial facies association (tidal channel and tidal bars). There is very well preserved thick regressive facies tract of regressive tide dominated delta. The regressive tide-dominated delta includes mosaic from two facies associations: 1- Fining upward and 2- Coarsening upward. Fining upward facies association represent tidal bar within tide dominated delta which consists of cross stratified sandstone with mudstone drapes. Coarsening upward facies association can be subdivided into three facies: 1- Bioturbated heterolithic sandstone, 2-Homogenized bioturbated sandstone and 3- Coarse grained lensoidal heterolithic. The maximum flooding interval 9 (MFI 9) is represented by 8 m (S20) of sandstones with moderately to intensively bioturbated branched network *Ophiomorpha* and going up to intensively bioturbated fossiliferous sandstone and the bioclastic fragments represented by e.g. bivalves, gastropods, echinodermata. This interval is also represented by 2.5 m (S9) of fossiliferous sandstone with echinodermata and gastropods or this 2.5 m could be part of a regressive facies tract. Sequence 9 is less widespread especially across much of the western study area with great variability and thickness (present in 9, 20, 21 and 22 sections). In the eastern areas it is not at all preserved. Sequence 9 is dated with Sr-isotopic data from 18.2 to 17.4 Ma from samples in the middle and the near from the top of the sequence.

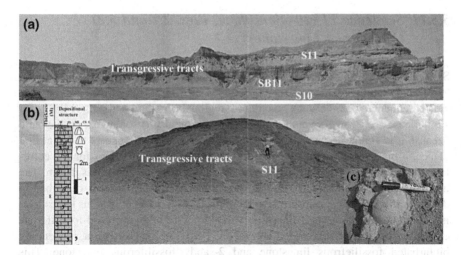

Fig. 5.16 (a) Photomosaic showing slope forming sandstone of S10, the transgressive facies tract consists of cross-bedded sandstone represented by transgressive estuary deposits. (b) Original photomosaic showing S11 that represented open shelf deposits which consists of bioturbated-fossiliferous carbonate. (c) Close view for this highly fossiliferous limestone including *scutella*

5.5.10 Sequence 10 (20–50 m thick; Figs. 2.1 and 5.2)

Sequence 10 is about 50 m thick (section 21) and reduces to 20 m in sections 9 & 20. This sequence is bounded by the SB10 below and by SB11 (regional maximum flooding surface) above (Fig. 5.16a), and consists of a very thick transgressive facies tracts. The regressive facies tract is not preserved. The transgressive facies tract consists of cross-bedded sandstone represented by transgressive estuary deposits. The transgressive estuary includes axial facies association (tidal channel and tidal bars). This sequence is similar to S9 has a less widespread extent especially across much of the western study area with great variability and thickness (present in 9, 20 and 21 sections). In the eastern areas it is not at all preserved. There are no regressive deposits preserved in Sequence 10 because the overlying sequence boundary 11 cuts down severely to truncate the entire upper part of this sequence, as it does in many of the sequences in the study succession. The maximum flooding interval 10 (MFI 10) is not easily to recognize in any section because it could be merged with the sequence boundary (SB10). This maximum interval is similar to the maximum flooding interval 3. This sequence lacks age control because the general lack of in-place fossils severely limits chronostratigraphic control of this sequence.

5.5.11 Sequence 11 (20–40 m thick, Figs. 2.1 and 5.2) (Marmarica Limestone)

Sequence 11 (Marmarica limestone) is about 40 m thick (section 32) and reduces to 20 m in section 9. This sequence is bounded by SB11 (This surface proved to be either sequence boundary or maximum regional flooding surface (MFS) (for similar case see Browning et al. 2006). So this sequence is somewhat problematic because it is similar to sequence 4. The sequence boundary appears to be, at first glance, a maximum flooding surface that is used as second datum for the correlation panel. However, SB11 is likely to be a short distance above this datum. The upper sequence boundary SB12 is not preserved. One facies tract is described in detail for this sequence: transgressive facies tract (Fig. 5.16a & b). This facies tract is interpreted as open shelf deposits which consists of bioturbated-fossiliferous carbonate that can be subdivided into facies, 1-fossiliferous limestion, moderately bioturbated fossiliefrous limestone and 2-sandy fossiliferous limestone. This highly fossiliferous limestone is rich in echinodermata (Fig. 5.16c), shell fragments, molds of pelecypod and gastropod. The base of this tract is lined by coarse grained, pebbles and mudclasts. Regressive facies tracts are lacking, either because regression was non- depositional or because regressive deposits were eroded during the final phase of regression and/or during the next transgression. The maximum flooding interval 11 (MFI 11) is usually not easily recognized. This interval occurs within open shelf deposits so it so hard to determine where the deepest level will be. Sequence 11 is dated with Sr-isotopic data from 16.5 to 17 Ma from samples in the middle and the near from the top of the sequence. The older ages obtained from the sequence probably result from either diagenesis or reworking of older material.

5.6 Discussion

A sequence stratigraphic interpretation has been presented for the lower Miocene Moghra Formation in the Qattara Depression of northwestern Egypt, based upon 18 measured sections and the mapping of erosional surfaces in superb desert outcrop over a lateral distance of more than 30 km. The Moghra Formation, which is exposed in a series of south-facing escarpments, consists of ∼200 m of estuarine, shelf and deltaic sandstones, siltstones, mudstones, and minor limestone, arranged into a series of transgressive–regressive cycles, each bounded by subregional erosion surfaces with <15 m of local erosional relief. These surfaces are partially armored by thin ferruginous pebble conglomerates containing petrified wood and bone fragments. The cycles, reaching thicknesses of up to 60 m each, are dominated by transgressive deposits, but many of them also have a thin capping of regressive shelf-to-deltaic deposits. The lower part of each cycle consists of stacked sets of unconsolidated cross-stratified sandstone and sandstone with locally abundant vertebrate fossils, petrified logs, *Thalassinoides* and *Ophiomor-pha* burrows. These deposits, which are interpreted as inner to outer estuarine, backstepping tidal channel-bar complexes, are capped by pervasively bioturbated *Ophiomorpha*-bearing sandstone beds with marine fauna, representing outermost estuary and marine shelf deposits. In some cycles, the estuarine beds pass upwards into a thin interval of crudely upward-coarsening shale and siltstone, representing regressive shelf-to-deltaic deposits but such deposits are often poorly preserved because they are commonly truncated by the next master erosion surface (sequence boundary). The upward change from transgressive estuarine deposits to regressive deltaic deposits necessarily passes up through a transgressive–regressive turn-around interval, known as the maximum flooding interval, represented by the most basin ward facies, the shelf deposits. Taken together, the series of Moghra sequences exhibit an overall transgressive trend, culminating in the open marine Marmarica Limestone. The transgressive tract typically has a basal tidal-fluvial interval, then an interval of estuarine tidal bars bounded beneath by a tidal rav-inement surface (Rt) and sometimes above by a wave ravinement surface (Rw), before transitioning upwards to open marine shelf deposits and a MFS recording maximum transgression. The regressive systems tract, usually a prograding tide-dominated deltaic unit (prodelta to delta-front deposits), is bounded beneath by MFS, and above by the next sequence boundary. Provisional ages, based on a combination of biostratigraphy and strontium isotope stratigraphy, range from 21 to 17 Ma (Table 5.1). The occurrence of some 11 transgressive–regressive cycles during this 4my interval strongly suggests that the cycles represent on average some 200–400ky, i.e., 4th-order sequences. The deep-marine oxygen isotope record for this interval is highly cyclic, with a subtle shift (to less positive δ^{18}O values) that is consistent a small sea-level rise over the 4my period. However, regional subsidence is required to account for most of the observed stratigraphic thickness. We infer that the preferential preservation of transgressive half cycles is due to the filling of available accommodation during 4th-order sea-level rises, alternating with erosive bypass of sediment to the Mediterranean margin during sea-level falls. The poorly preserved shelf-to-deltaic half cycles represent reduced rates of sea-level rise (or increased rates of sediment supply) prior to each sea level fall. The common absence or thinness of the regressive part of cycles is due to (1) the severity of erosion caused by relative sea-level fall, prior to subsequent sea-level rise and renewed transgression, (2) to sea-level fall occurring before the completion of transgression, i.e. the erosional valley was not entirely infilled before a new fall occurs, or (3) to the relatively proximal setting of the study area, such that the regressive half-cycle is not properly developed, but is rather repre-sented by mainly erosion and sediment bypass. In this last option the better developed regressive tracts are predicted to occur downdip from the study area, i.e. farther basin ward where more symmetrical cycles will occur. No lowstand sys-tems tracts or major downshift surfaces have been recognized within the succes-sion. The Moghra outcrops are significant as a superb archive of fluvially fed, marginal-marine reservoir sands, and as a reference for the timing of probably eustatically generated sequence boundaries in a tectonically simple setting far from the North Atlantic region. New drilling at the New Jersey continental margin (IODP Expedition 313, 2009) will focus on the same lower Miocene interval.

5.7 Tectonic Subsidence and Accommodation

The other important objective of this study is to quantitatively evaluate the effects of eustasy, tectonics, and sediment supply variations on Early Miocene sequences in the Qattara Depression, North Western Desert, Egypt. What is the origin of the space in which these thick sediments accumulated?. Unsurprisingly, it is likely, as in most successions that subsidence is the key, particularly because the eustatic sea-level rise over the interval of interest (Early Miocene) is likely to have been relatively small. It is widely accepted that sediment supply and accommodation space are the dominant controls on stratigraphic packaging. Sea-level rise or subsidence controls the creation of accommodation space. During the period recorded in the study succession (Early Miocene), the eustatic rise recorded on the global curve is less than 10 meters (Pekar et al. 2002, 2006), suggesting that subsidence was the major factor in creating accommodation space. The most characteristic feature of the studied succession is the regular repetition of 2–400 ky erosively-bounded sequences The subsidence of the basin has been calculated, showing that a period of strong to weak (>60 m/my) subsidence occurred during Early Miocene when the cyclic succession developed. The tectonic subsidence would explain the aggradational stratal pattern of the complete succession, the lack of emersion features with high sedimentation rate in shallow marine waters (a succession 230 m thick was deposited at most in 4 my). The significance of that conclusion is that the *overall* deepening at Moghra cannot be ascribed only to sea-level rise, because much of the space being created is relative sea-level rise created by tectonic subsidence.

References

Bosellini A (1984) Progradation geometries of carbonate platforms: examples from the Triassic of the Dolomites, northern Italy. Sedimentology 31:1–24

Bosellini A (1988) Outcrop models for seismic stratigraphy: examples from the Triassic of the Dolomites. In: Bally AW (ed) Atlas of seismic stratigraphy, AAPG studies in geology, p 194–203

Browning J, Miller K, Mclaughlin P, Kominz M, Sugarman P, Monteverde D, Feigenson M, Hernández J (2006) Quantification of the effects of eustasy, subsidence, and sediment supply on Miocene sequences, mid-Atlantic margin of the United States. Geol Soc Am Bull 118:567–588

Catuneanu O, Abreu V, Bhattacharya J, Blum M, Dalrymple R, Eriksson P, Fielding C, Fisher W, Galloway W, Gibling M (2009) Towards the standardization of sequence stratigraphy. Earth-Sci Rev 92:1–33

Christie-Blick N (1991) Onlap, offlap, and the origin of unconformity-bounded depositional sequences. Mar Geol 97:35–56

Christie-Blick N, Driscoll N (1995) Sequence stratigraphy. Annu Rev Earth Planet Sci 23:451–478

Christie-Blick N, Grotzinger J, Von Der Borch C (1988) Sequence stratigraphy in proterozoic successions. Geology 16:100–104

Dalrymple RW (1992) Tidal depositional systems. In: Walker RG, James NP (eds) Facies models: response to sea level change. Geological Association of Canada, St John's, pp 195–218

De Graciansky P, Hardenbol J, Jacquin T, Vail P (1998) Mesozoic and cenozoic sequence stratigraphy of European basins, v. Special Publication 60, Society for Sedimentary Geology, p 786

García-Mondéjar J, Fernández-Mendiola PA (1989) Evolución plataforma/cuenca en el Albiense de Lunada y Soba (Burgos y Cantabria). Secuencias, asociaciones de sistemas sedimentarios (systems tracts) y cambios del nivel del mar. XII Congr. Esp. Sedimentol., Libro-Guía Excursiones Geol., Excursión No. I. Dep. Estratigr., Geodin. Paleontol. (Univ. País Vasco), and Div. Invest. Recursos (Ente Vasco de la Energía), Leioa-Bilbao.7–43

García-Mondéjar J, Fernández-Mendiola PA (1991) Sequence stratigraphy and systems tracts of a mixed carbonate and siliciclastic platform-basin model: The Albian of Lunada and Soba, northern Spain: Am. Assoc. Pet. Geol. in prep

Haq B, Hardenbol J, Vail P, (1987) Chronology of fluctuating sea levels since the Triassic. Science 235:1156

Helland-Hansen W (2009) Towards the standardization of sequence stratigraphy. Earth-Sci Rev 94:95–97

Kidwell S (1997) Anatomy of extremely thin marine sequences landward of a passive-margin hinge zone. J Sediment Res Sect A and B 67:322–340

Locker S, Hine A, Tedesco L, Shinn E (1996) Magnitude and timing of episodic sea-level rise during the last deglaciation. Geology 24:827–830

Miall A (1995) Whither stratigraphy? Sed Geol 100:5–20

Miller K, Sugarman P, Browning J, Kominz M, Olsson R, Feigenson M, Hernández J (2004) Upper Cretaceous sequences and sea-level history, New Jersey coastal plain. Geol Soc Am Bulletin 116:368–393

Mitchum Jr R, Vail P, Thompson III S (1977) Seismic stratigraphy and global changes of sea level, Part 2. The depositional sequence as a basic unit for stratigraphic analysis. In: Payton CE (ed) Seismic stratigraphy—applications to hydrocarbon exploration. American Association of Petroleum Geologists, Memoir, p 53–62

Muto T, Steel R (1997) Principles of regression and transgression: the nature of the interplay between accommodation and sediment supply. J Sediment Res Sect a and B 67:994–1000

Muto T, Steel R, Swenson J (2007) Autostratigraphy: a framework norm for genetic stratigraphy. J Sediment Res 77:2–12

Mutti E, Rosell J, Allen G, Fonnesu F, Sgavetti M (1985) The Eocene Baronia tide dominated delta-shelf system in the Ager Basin. In: Mila MD, Rosell J (eds) Excursion Guidebook, 6th European Regional Meeting: Spain. International Association of Sedimentologists, Lleida, pp 579–600

Pekar S, Christie-Blick N, Kominz M, Miller K (2002) Calibration between eustatic estimates from backstripping and oxygen isotopic records for the Oligocene. Geology 30:903

Pekar S, Deconto R, Harwood D (2006) Resolving a late Oligocene conundrum. Deep-sea warming and Antarctic glaciation. Palaeogeogr Palaeoclimatol Palaeoecol 231:29–40

Posamentier H, Vail P (1988) Eustatic controls on clastic deposition II—sequence and systems tract models. In: Wilgus CK, Hastings BS, Kendall CGStC, Posamentier HW, Ross CA, Van Wagoner JC (eds) Sea-level changes: an integrated approach, SEPM Special Publication, pp 125–154

Posamentier H, Jervey M, Vail P (1988) Eustatic controls on clastic deposition I—conceptual framework. Sea level changes: an integrated approach, vol 42. SEPM Special Publication, pp 109–124

Reynolds D, Steckler M, Coakley B (1991) The role of the sediment load in sequence stratigraphy. The influence of flexural isostasy and compaction. J Geophys Res 96:6931–6949

Sarg J (1988) Carbonate sequence stratigraphy. In: Wilgus CK, Hastings BS, Kendall CGSC, Posamentier HW, Ross CA, Van Wagoner JC (eds) Sea-level changes: an integrated approach. Special Publication, SEPM, pp 155–181

Sarg J (1989) Middle-late Permian depositional sequences, Permian basin, west Texas and New Mexico. In: Bally AW (ed) Atlas of seismic stratigraphy AAPG studies in geology, pp 140–154

Siggerud E, Steel R (1999) Architecture and trace-fossil characteristics of a 10,000–20,000 year, fluvial-to-marine sequence, SE Ebro Basin, Spain. J Sediment Res 69:365–383

Vail P (1987) Seismic stratigraphy interpretation using sequence stratigraphy. Part 1: seismic stratigraphy interpretation procedure. In: Bally AW (ed) Atlas of seismic stratigraphy, American Association of Petroleum Geologists Geology pp 1–10

Vail P, Mitchum Jr R, Thompson III S (1977) Seismic stratigraphy and global changes of sea level, part 3: Relative changes of sea level from coastal onlap. In: Payton CE (ed) Seismic stratigraphy—applications to hydrocarbon exploration, American Association of Petroleum Geologists, pp 63–81

Van Wagoner J, Posamentier H, Mitchum R, Vail P, Sarg J, Loutit T, Hardenbol J (1988) An overview of the fundamentals of sequence stratigraphy and key definitions. In: Wilgus C, Hastings BS, St. C. Kendall CG, Posamentier HW, Ross CA, Van Wagoner JC (eds) Sea-level changes-an integrated approach Society of Economic Paleontologists and Mineralogists (SEPM), Special Publication, pp 39–45

Van Wagoner J, Mitchum R, Campion K, Rahmanian V (1990) Siliciclastic sequence stratigraphy in well logs, cores, and outcrops. Concepts for high-resolution correlation of time and facies. AAPG, Methods in Exploration Series, vol 7, p 55

Wilgus C, Hastings B, Kendall C, Posamentier H, Ross C, Van Wagoner J (1988) Sea-level changes: an integrated approach. Society of Economic Paleontologists and Mineralogists Special Publication, vol 42, p 407

Chapter 6
Model for Development of Moghra Estuarine Complex

Abstract The net transgressive Lower Miocene Moghra Formation of Egypt is a sandy estuarine complex consisting of a series of stratigraphic units that reflect repeated transgressive to regressive shoreline movements across the Burdigalian (Lower Miocene) coastal landscape. The transgressive part of each unit is preserved atop a deep erosional scour surface, and consists of tidal–fluvial sandstones with tree logs and vertebrate bones that transition up to cross-stratified, tidal estuarine channel deposits and then to open-marine, shelf mudstones and limestones. In contrast, the regressive part is thinly developed and consists of thin-bedded, fossiliferous shelf mudstones that pass upward to thin, tide-influenced delta-front deposits. Each of the nine transgressive–regressive units of the Moghra Formation is capped by a river-scour surface that severely truncates the underlying regressive half-unit. Two main alternative models are suggested for Moghra Formation: (A) Erosion of a large (20–30 km wide, 20–60 m deep) master valley during a major Early Miocene sea-level fall, then infilling by repeated (4th-order time scale) transgressive–regressive cycles driven by (1) eustatic SL falls or by (2) autocyclicity, each during steady transgression into the valley and steady sediment supply; or by (B) subsiding (steady sea-level rise) tectonic depression (no large valley) infilled by T-R cycles. The T-R cycles were driven by (1) eustatic sea-level falls, by (2) short tectonic uplifts (to create updip erosion and downdip deltas as erosion products), or by (3) climate-driven sediment supply pulses (this would also contain some autogenic lobe shifting that would not be depression-wide). We prefer the climate-driven sediment supply pulses because we see no evidence of a large eroded valley, because eustatic falls would have been modest in early Miocene, and because there is no evidence of repeated regional tectonic uplifts to create the relative sea level falls. Climate changes are preferred because they are known to happen at high frequency and they are a major cause of sediment flux changes to basins.

S. M. Hassan, *Sequence Stratigraphy of the Lower Miocene Moghra Formation in the Qattara Depression, North Western Desert, Egypt*, SpringerBriefs in Earth Sciences, DOI: 10.1007/978-3-319-00330-6_6, © The Author(s) 2013

6.1 Model for Moghra Formation

Sedimentological and sequence stratigraphic analysis highlights and raises many questions about a complex interaction between sediment supply, glacio-eustatic sea-level change, regional uplift, and local tectonics, providing insight into the main relationships between sequence architecture and the above external controls. In addition, some of the architectural changes are likely to be autogenic in origin, i.e. stratigraphic responses without any rate changes in any of the external driving variables. The aim of this chapter is to shed light on two alternative models for Moghra Formation. This is the first study to use this integrated approach to unravel the depositional setting of the Moghra Formation.

Basically, the two main alternative models are: (A) Erosion of a large (20–30 km wide, 20–60 m deep) master valley during a major Early Miocene sea-level fall, then infilling by repeated (4th-order time scale) transgressive–regressive cycles driven by (1) eustatic SL falls or by (2) autocyclicity, each during steady transgression into the valley and steady sediment supply; or by (B) subsiding (steady sea-level rise) tectonic depression (no large valley) infilled by T-R cycles driven by (1) eustatic sea-level falls, or by (2) short tectonic uplifts (to create updip erosion and downdip deltas as erosion products), or by (3) climate-driven sediment supply pulses (this would also contain some autogenic lobe shifting that would not be depression-wide). We prefer B (3), because we see no evidence of a large eroded valley, because eustatic falls would have been modest in early Miocene, and because there is no evidence of repeated regional tectonic uplifts to create the relative sea level falls. Climate changes are preferred because they are known to happen at high frequency and they are a major cause of sediment flux changes to basins. B3 implies a steady or increasing, long-term, subsidence-driven rise of base level, during which tide-dominated estuary development was punctuated repeatedly by sediment supply pulses and consequent appearance of brief but significant regressive delta growth. In this way, estuary-delta couplets were generated in the Moghra succession. A discussion 'for' and 'against' each of the above hypotheses is made below.

(A1) erosion of a large valley then infilling by repeated transgressive–regressive cycles driven by eustatic SL changes. This hypothesis requires an initial large eustatic SL fall to erode the valley then repeated rise-fall cycles thereafter to infill

For—The lower Miocene Moghra Formation in the Qattara Depression consists of ~230 m of fluvially dominated estuarine and marginal marine mixed silici-clastic/carbonate deposits, arranged into a series of transgressive–regressive cycles bounded by regional erosion surfaces with <15 m of local erosional relief. The cycles, reaching thicknesses of up to 60 m each, are dominated by transgressive deposits and the depth of the individual incised valleys range from 30 to 60 m. The measured sections within this valley cover more than 35 km. The lower part of each cycle consists of unconsolidated cross-stratified sandstone with locally abundant vertebrate fossils, petrified logs, and *Thalassinoides* and *Ophiomorpha*

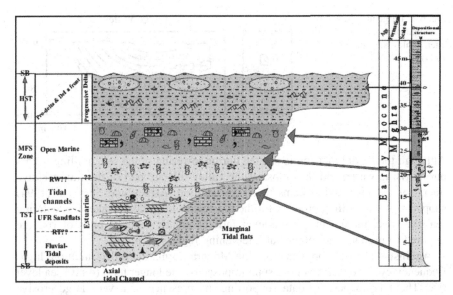

Fig. 6.1 Diagrammatic cross section showing distribution of facies and depositional environment within early Miocene-estuarine incised valley of Moghra Formation, with the inferred sequence-stratigraphic interpretation

burrows (Fig. 6.1). These deposits, which are interpreted as a backstepping estuarine channel complex, are capped by pervasively bioturbated *Ophiomorpha*-bearing sandstone beds with an open marine fauna. All these criteria could possibly prove that Moghra is a large incised-valley filled by repeated transgressive–regressive cycles driven by unsteady eustatic SL change. The absence of lowstand deposits in the incised valley can be interpreted in terms of lowstand acting as a bypass zone, or lowstand deposits were eroded and reworked during the subsequent transgression.

Ashley and Sheridan (1994) they developed depositional models for valley-fill sequences formed on a passive continental margin which it's lower bounding unconformity (SB I) is produced by fluvial incision during sea-level lowstands. In addition, they suggest the large "drowned-river" valleys are characterized by a sediment sandwich of sand-mud-sand which may contain several marine erosion surfaces and have a high proportion of sand. In addition, during transgressions, depth of erosion is estimated at 5–10 m, and thus preservation potential for the lower portion of the valley fill is excellent for the large valleys (Fig. 6.2). Furthermore, topography is an important factor in valley-fill preservation on passive margins. Depressions created from fluvial or shoreface erosion often provide the only "accommodation space" on continental shelves of passive margins that will shield transgressive tract deposits from erosion and ensure preservation of the geological record.

Consequently, the Moghra Formation can be considered as large incised valley in an analogous manner to the hypotheses of Ashley and Sheridan (1994).

Fig. 6.2 Large incised valley fill might originate (Modified after Ashley and Sheridan 1994)

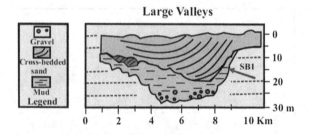

Although the fluvial incision is created during lowering of the sea, infilling occurs during rising water, and thus most of the valley fills are part of the transgressive tract, not the lowstand systems tract). In addition, Ashley and Sheridan (1994) support that the infilling process, for the most part of the valley, takes place during sea-level rise, though minor sea-level falls or greatly reduced rates of rise are required to produce the minor but intervening regressive pulses .

Against—The outcrop data from the Moghra Formation reflect multiple high-frequency cycles of sea level fall superimposed on the longer term (4 my) sea level rise. There is no outcrop data supporting the presence of a very large erosive master valley. In terms of the global sea-level curve of Haq et al. (1987, 1988), the time-span is characterized by a sharp rise and little fluctuation in sea level of third order cycles 2.1 and 2.2 (upper part) (see Fig. 6.3, modified from Berggren et al. 1995a, look for the area of interest). This evidence may possibly be consistent with the 10 or 11 high-frequency Moghra transgressive—regressive cycles superimposed on a long-term sea level rise during the same period (Berggren et al. 1995a), but it is not consistent with a very large initial fall of sea level.

Another question arising concerns the origin of the space in which the Moghra sediments accumulated? Subsidence is likely to be the key, particularly during the period of the study succession (early Miocene). The eustatic rise recorded on the global curve around 40–50 m (Abreu and Anderson 1998 see Fig. 6.4, look the area of interest) or is less than 10 m (Pekar and Christie-Blick 2008 see Fig. 6.5, look the area of interest), suggesting that subsidence was the major factor in creating accommodation space, and not eustatic sea level change. The significance of that conclusion is that the *overall* deepening at Moghra cannot therefore be ascribed to eustatic sea-level rise. Had the rate of subsidence been just a little less or the sediment supply just a little higher, then we would observe overall shoaling/regression.

Furthermore, two types of high-frequency cyclical isostatic adjustments should be considered in incised-valley models Blum (2008). First, the magnitude of sediment removed and replaced in large incised-valley systems over the course of a glacio-eustatic cycle produces a corresponding cycle of flexural uplift and subsidence that enhances depths of incision during the lowstand, and the overall thickness of transgressive to highstand strata. Rates of uplift or subsidence due to sediment unloading and loading may exceed long-term rates of accommodation change by an order of magnitude, and affect stratal geometries for distances of 10–100 s of kilometers from the depocenter. Second, sea-level fall and removal of water from a broad shelf should result in hydro-isostatic uplift, with the reverse process during

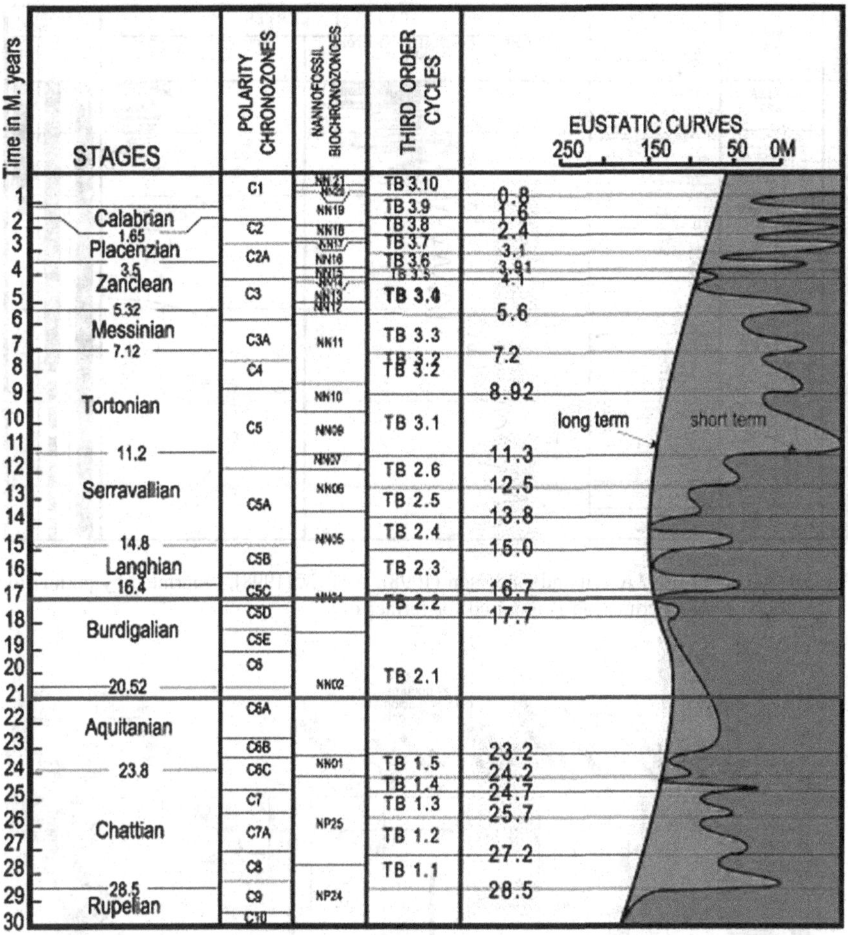

Fig. 6.3 Based on Haq et al. (1987) and Beggren et al. (1995a)

transgression and highstand. River systems are obliged to extend across the shelf during sea-level fall, but they need not cut valleys that are any deeper than a single channel belt: hydro-isostatic uplift may be the driving force behind the deeper incised valleys that are commonly observed Blum (2008). Therefore, we haven't sure evidence for these flexural uplift pulses in Moghra; however, we have little evidence for it but, we put it tentatively. Similarly, we have no evidence that the uplift may be the driving force behind the deeper incised valleys.

(A2) **erosion of a large valley then infilling by repeated transgressive–regressive cycles driven by autocyclicity during steady sea-level rise and strong steady supply,**

For—For more understanding, we need to know the term autogenic. As used here according to Muto et al. (2007) autogenic refers to a stratigraphic response

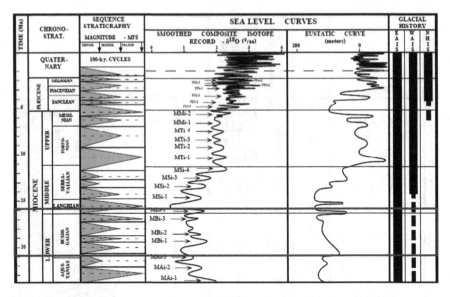

Fig. 6.4 Modified from Abreu and Anderson (1998). AAPG©[1998], "reprinted by permission of the AAPG whose permission is required for further use".

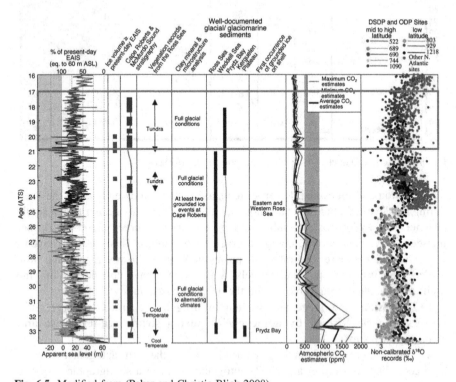

Fig. 6.5 Modified from (Pekar and Christie-Blick 2008)

that appears despite steady external forcing (Blum and Törnqvist 2000; Muto and Steel 2004, 2001; Muto and Swenson 2006; Swenson et al. 2005). Allogenic, in contrast, refers to stratigraphic responses that appear as a result of unsteady external forcing (Muto and Swenson 2005a, 2005b). For example, a sedimentation change in response to accelerating sea-level rise can be regarded as an allogenic process (with respect to sea level change), whereas an architectural change occurring during constant rate of sea level rise is regarded as arising from autogenic process. It is important to note that "steady sea-level forcing" means forcing by sea level that is changing at a constant rate. In standard sedimentologic usage, an autogenic response arises from internal feedback between transport processes, happening during constant external forcing. Common examples include river avulsion and delta-lobe switching (Beerbower 1965; Correggiari et al. 2005; Hansen and Rasmussen 2008; Mckeown 2004; Miall 1996), which can be regarded as local or stochastic processes and can occur multiple times under constant forcing of the depositional system. Though the usage of "autogenic" here is essentially unchanged from classic sedimentological usage, the autogenic processes that we treat here are mostly those that operate over the entire system. They are deterministic, and occur only a single time during the entire period of a given steady forcing. Our use of "autogenic" does not imply a particular, absolute spatio-temporal scale, and it is possible that the time scale for the intrinsic non-equilibrium response of the fluvio-deltaic system overlaps with the time scales of other autogenic responses Muto et al., (2007).

Against—It is simply not possible to create repeated T-R sequences by autogenic means! Autoretreat can be produced one time, but subsequent regression cannot be achieved without a change in the external forcing, either by increase in sediment supply or by fall of SL/tectonic uplift. In addition, these autocyclic processes take place on relatively short time scales, typically hundreds to thousands of years, and should be considered the potential control on the short-term sequences (Laurin and Sageman 2007). Therefore, it difficult to explain it in terms of autocyclic processes.

(B1) Subsiding Tectonic Depression (no large valley) steadily infilled by T-R cycles driven by eustasy (Fig. 6.6),

For—Tectonically driven subsidence is required to account for the thickness of accumulated sediment. The very small amplitudes of eustatic sea-level rise discussed above for the early Miocene are insufficient to produce the observed stratigraphy.

Fig. 6.6 Modified from Laurini and Sageman (2007)

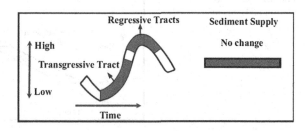

The sequences in Moghra are stacked slightly retrogradationally (landwards) as a result of overall tectonic subsidence being greater than supply. In order for the regressive punctuations to occur between transgressions, significant (greater than the subsidence rate) eustatic falls would have had to occur.

Against—There is no field evidence of forced regression and there is no theoretical basis for eustatic falls in this time interval being more than 10–20 m. This suggests strongly that eustasy was not the prime drive to these transgressive–regressive cycles that accumulated to a maximum thickness of 230 m.

(B2) Steadily subsiding tectonic depression (no large valley) infilled by T-R cycles driven by short tectonic uplifts (to create erosion and downdip deltas as erosion products).

For—Focusing on the Moghra sequences, we will find that 6 or 7 sequences are truncated without preservation for any regressive deposits. The others sequences include thin regressive deposits and only few sequences have thick regressive deposits. This may be possible to explain in terms of short tectonic uplift pulses (not continuous uplift just pulses) resulting in non-deposition or erosion of these tracts. Moreover, microscopic study establish that the poor maturity of the mixed-interval deposits (textural immaturity reflected by angular grains, poor sorting, and compositional immaturity represented by predominant rock fragments from chert, other fragments from igneous and metamorphic rocks and detrital minerals from feldspars and plagioclase) is consistent with the idea of a short-time sedimentary cycle for the metamorphic grains or igneous rocks and this can give evidence the transgressive–regressive cycles driven by short tectonic uplifts pulses.

Against—There is no evidence of such frequent tectonic uplifts, and varying rate of subsidence alone would not produce the regressive half cycles.

(B3) Steady subsiding tectonic depression (no large valley) infilled by T-R cycles driven by climate-induced sediment supply pulses (this should also contain some autogenic lobe shifting that would not be depression-wide.

For—The T-R cycles in the Moghra succession, created during steady subsiding tectonic depression, could be driven by sediment supply pulses (Fig. 6.7).

Two sets of evidence are used to support this hypothesis. The two evidences will support the idea that the transgressive- regressive cycles are driven by the climate.

The first set of evidence uses climate changes; for example drier to wetter periods caused by changing monsoon intensity, are known to double or triple the

Fig. 6.7 Modified from Laurini and Sageman (2007)

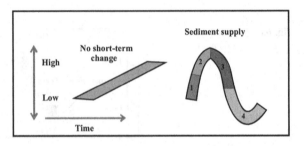

volume of sediment yielded into the basin. Goodbred (2003) who concluded that (1) sedimentary signals in the Ganges are transferred rapidly from source to sink with little apparent attenuation, and (2) that this very large dispersal system responds to multimillennial-scale climate change in a system-wide and contemporaneous manner. It is suggested that these responses are a function of the summer monsoon and its overarching control of the system's hydrology and, hence, sediment production and transport. In addition, he suggested that the Ganges dispersal system, probably more than most, is dominated by climatic forcing, thereby making it possible to faithfully recognize climate-induced responses. Similar patterns might be most appropriately sought in other highly seasonal, monsoon-forced dispersal system typical of the low latitudes (e.g., Prins et al. 2000; Thomas and Thorp 1995). Similarly we can expect this scenario in Moghra and with the monsoon to play a role to increase the amount of sediment supply that was transported to the basin.

The second set of evidence supports the climate driving force in Moghra Formation in terms of the the transgressive–regressive cycles that dominated in Moghra Formation, consisting of mixed siliciclastic—carbonate phases interpreted to record transition zone between subtropical and temperate conditions, characterized by: (1) heavy discharges of coarse-grained siliciclastic sediments; (2) molechfor carbonate factories which the molechfor lithofcaies, consisting of benthic foraminifers, molluscus, echinoids, bryozoans and barnacles. These carbonate lithofacies present complex distribution patterns seemingly related primarily to latitude and depth that control water temperature, although other factors (e.g., River discharge, water circulation and suspended sediment) controlling water salinity and temperature, nutrient content, light penetration also play important roles (for same lithofacies see Carannante et al. 1988); and (3) high-energy hydrodynamic regimes with intense persistent currents flowing towards the inner shelf. On the basis of the sedimentary structures, the hydrodynamic conditions during deposition of the mixed siliciclastic—carbonate interval were high-energy. Based on sedimentary and biogenic features of the transgressive–regressive cycles and their temporal durations (ca. 200–400 ka), the cycles that include the mixed siliciclastics—carbonate is interpreted to reflect warm-cool episode driven by precession orbital cycles. The correlation with orbital cycles revealed that individual mixed siliciclastic—carbonate correspond to eccentricity maxima at 100–400 ka, respectively (Berggren et al. 1995b to see the duration of eccentricity maxima). The diagram (Fig. 6.8) which modified of García-García et al. (2009) is showing that during warm- subtropical that may be continue to wet-temperate episodes, the shelf is subject to intense siliciclastic supply due to the increase in activity of the feeder systems. Simultaneously, typical warm to cool-water carbonate factories arise. Both the siliciclastic and the skeletal components are dispersed towards the basin. Therefore, we can conclude that the climate can affect on the amount of sediment supply that have driven to the basin. On the other hand the scenario will be different during the highstand sequences which indicate that increased progradation and less great mixing of shelf carbonates with nearshore

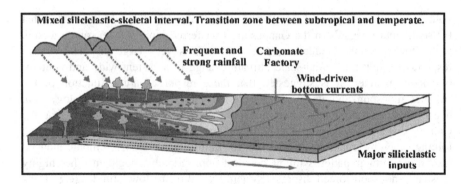

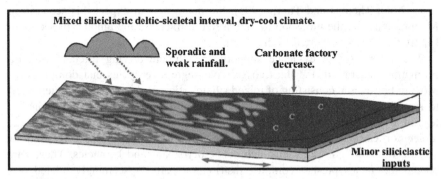

Fig. 6.8 Modified from García-García et al. (2009)

siliciclastics, likely in response to lowering sea-levels and cooling climate during the next sea level associated with the next sequence boundary.

Against—Against this hypothesis is the fact that there is no direct and clear evidence in Moghra succession that the climate repeatedly changed 11 times. We can hypothesize that the transgressive units always imply wet periods, but we have no strong evidence that the highstand or sequence boundary intervals necessarily reflect dry periods.

6.2 Why Moghra is Estuarine but not Within an Incised Valley?

Basically, there is simply no evidence for a single large master valley wall system that enclosed Moghra succession. According to Dalrymple (2006), estuaries (sensu Dalrymple et al. 1992) are indeed common within valleys, but the presence of a valley is not as fundamental and important a criterion as other aspects of the definition. Instead, the facies stacking pattern and the direction(s) of sand transport are much more important criteria. From a hydrodynamic point of view, the essential and basic aspects of a coastal system are the direction(s) of net sediment transport. Thus, a system that is flood dominated (i.e., there is a net import of sediment from the sea by means of waves and/or tidal currents-the system is "import dominated"), is an estuary. This distinction, in turn, is related to the existence of unfilled accommodation space landward of the coastline. The existence of unfilled space leads to the import of sediment from the sea. In addition, the creation of the unfiled space is associated with relative sea-level rise and transgression. So, the co-occurrence of landward sediment transport and transgression is a fundamental association that is restricted to estuaries. Dalrymple (2006) suggests that unfilled space exists within drowned river valleys, but unfilled space (and, by extension, landward transport of sediment and the existence estuaries) can also exist in other settings.

6.3 Preferred Model for the Cycles of the Moghra Fm

On the basis of the above discussion, we prefer a non-valleyed interpretation for the Moghra succession, mainly because there is no field evidence for master valley walls to enclose the entire succession. The alternative, the subsiding tectonic depression model, is therefore more consistent with the stratigraphic evidence. Regarding the cyclic (repeated transgressive–regressive) character of the entire infill, we reject high-frequency tectonic uplifts or eustatic sea-level falls to explain the repeated regressive half-cycles, again because there is little supportive evidence. On the other hand, with the overall subsidence context (causing overall relative sea level rise), the regressive half cycles can be simply and economically explained by climate-driven sediment supply increases. Increased sediment flux during subsidence-driven transgression is sufficient to cause a short-term regression turnaround in the stratigraphy. Nevertheless the transgressive half cycles (the estuaries) do dominate the succession, requiring only very short punctuations of increased supply to interrupt the dominant transgressive tendency.

References

Abreu VS, Anderson JB (1998) Glacial eustasy during the Cenozoic; sequence stratigraphic implications. AAPG Bull 82:1385–1400

Ashley G, Sheridan R (1994) Depositional model for valley fills on a passive continental margin: incised-valley systems: origin and sedimentary sequences, pp. 286–301, SEPM, Special Publication

Beerbower J (1965) Cyclothems and cyclic depositional mechanisms in alluvial plain sedimentation, pp. 31–42

Berggren WA, Hilgen FJ, Langereis CG, Kent DV, Obradovich JD, Raffi I, Raymo ME, Shackleton NJ (1995b) Late Neogene chronology: new perspectives in high-resolution stratigraphy. Geol Soc Am Bull 107:1272–1287

Blum M (2008) New views on incised valleys: insights from quaternary systems of the Gulf of Mexico coast and shelf: Geological Society London Meeting on Rivers, estuaries, deltas & beaches: Traps For Fossil Fuels, Abstract, p. 17

Blum M, Törnqvist T (2000) Fluvial responses to climate and sea-level change: a review and look forward. Sedimentology 47:2–48

Carannante G, Esteban M, Milliman JD, Simone L (1988) Carbonate lithofacies as paleolatitude indicators; problems and limitations. Sedment Geol 60:333–346

Correggiari A, Cattaneo A, Trincardi F (2005) The modern Po Delta: lobe switching and asymmetric prodelta growth. In: Trincardi F, Syvitski J (eds) Mediterranean Prodelta Systems, Proceedings of the international ComDelta symposia, p. 49–74, Aix-en-Provence: France, Marine Geology

Dalrymple RW (2006) Incised valleys in time and space; an introduction to the volume and an examination of the controls on valley formation and filling, vol. 85. Special Publication—Society for Sedimentary Geology, pp. 5–12

Dalrymple RW, Zaitlin BA, Boyd R (1992) Estuarine facies models: conceptual basis and stratigraphic implications. J Sediment Petrol 62:1130–1146

García-García F, Soria JM, Viseras C, Fernandez J (2009) High-frequency rhythmicity in a mixed siliciclastic-carbonate Shelf (Late Miocene, Guadix Basin, Spain): a Model of Interplay Between Climatic Oscillations, Subsidence, and Sediment Dispersal. J Sediment Res 79:302–315

Goodbred S (2003) Response of the Ganges dispersal system to climate change: a source-to-sink view since the last interstade. Sediment Geol 162:83–104

Hansen J, Rasmussen E (2008) Structural, sedimentologic, and sea-level controls on sand distribution in a steep-clinoform asymmetric wave-influenced delta: Miocene Billund Sand, Eastern Danish North Sea and Jylland. J Sediment Res 78:130–146

Haq B, Hardenbol J, Vail P (1987) Chronology of fluctuating sea levels since the Triassic. Science 235:1156

Haq B, Hardenbol J, Vail P (1988) Mesozoic and Cenozoic chronostratigraphy and cycles of sea-level change. In: Wilgus CK, Hastings BS, Kendall CGSC, Posamentier HW, Ross CA, Van Wagoner JC (eds) Sea-level changes: an integrated approach: society of economic paleontologists and mineralogists special publication, SEPM Special Publication, pp. 71–108

Laurin J, Sageman B (2007) Cenomanian turonian coastal record in SW Utah, USAOrbital-scale transgressive regressive events during Oceanic anoxic event II. J Sediment Res 77:731–756

Mckeown HA, Bart PJ, Anderson JB (2004) High-resolution stratigraphy of a sandy, ramp-type margin—Apalachicola, Florida, U.S.A. In: Anderson JB, Fillon RH (eds) Late quaternary stratigraphic evolution of the northern Gulf of Mexico Margin. SEPM, Special Publication, pp. 25–41

Miall A (1996) The geology of fluvial deposits; sedimentary facies, basin analysis, and petroleum geology. Springer, Berlin, p. 582

Muto T, Steel RJ (2001) Autostepping during the transgressive growth of deltas: results from flume experiments. Geology 29:771–774

Muto T, Steel R (2004) Autogenic response of fluvial deltas to steady sea-level fall: implications from flume-tank experiments. Geology 32:401–404

Muto T, Swenson J (2005a) Large-scale fluvial grade as a nonequilibrium state in linked depositional systems: theory and experiment. J Geophys Res 110(F): F03002 doi: 10.1029/2005JF000284

Muto T, Swenson J (2005b) Controls on alluvial aggradation and degradation during steady fall of relative sea level: flume experiments. In: Parker G, Garcia MH (eds) River coastal and estuarine morphodynamics. Talor and Francis, London, pp 345–367

Muto T, Swenson J (2006) Autogenic attainment of large-scale alluvial grade with steady sea level fall: an analog tank/flume experiment. Geology 34:161–164

Muto T, Steel R, Swenson J (2007) Autostratigraphy: a framework norm for genetic stratigraphy. J Sediment Res 77:2–12

Pekar S, Christie-Blick N (2008) Resolving apparent conflicts between oceanographic and Antarctic climate records and evidence for a decrease in pCO2 during the Oligocene through early Miocene (34–16 Ma). Palaeogeogr Palaeoclimatol Palaeoecol 260:41–49

Prins MA, Postma G, Cleveringa J, Cramp A, Kenyon NH (2000) Controls on terrigenous sediment supply to the Arabian Sea during the late Quaternary; the Indus Fan. Mar Geol 169:327–349

Swenson J, Paola C, Pratson L, Voller V, Murray A (2005) Fluvial and marine controls on combined subaerial and subaqueous delta progradation: morphodynamic modeling of compound-clinoform development. J Geophys Res 110 (F)(F03007):16. doi: 10.1029/2004JF000208

Berggren WA, Kent DV, Swisher I, CC, and Aubry M-P (1995a) A revised Cenozoic Geochronology and Chronostratigraphy. In: Berggren WA, Kent DV, Aubry M-P, Hardenbol J (eds) Geochronology time scales and global stratigraphic correlation SEPM, Special Publication

Thomas MF, Thorp MB (1995) Geomorphic response to rapid climatic and hydrologic change during the late Pleistocene and early Holocene in the humid and sub-humid tropics. Quatern Sci Rev 14:193–207

Chapter 7
Conclusion

The northern cliffs of the Qattara Depression exhibit excellent outcrops of the Lower Miocene Moghra Formation, which is known for its fossil vertebrates. Despite this successful focus on the vertebrate of Moghra area, there is still a noticeable gap in our knowledge about Lower Miocene sedimentology and sequence stratigraphy of this area. The literature about the sedimentology and petroleum geology of this part of the Moghra Formation has been sparse and slow in coming, despite some excellent work over the years by academic and petroleum workers in this area. Thus, great attention is paid in this present study to establish a robust sedimentology and high-resolution sequence stratigraphic framework for the Lower Miocene Moghra Formation. Included are works based on outcrop study and, most importantly, new sedimentological and stratigraphic information not previously available. Moreover, the palaeogeographic reconstruction and facies distribution of the Lower Miocene in this area (onshore Mediterranean) will directly impact the offshore petroleum exploration strategies in terms of reservoir prediction and proximal-distal facies variation.

7.1 Facies Associations

The deposits of the Moghra formations have been classified into twenty facies and grouped into ten facies associations. See Table 7.1.

Depositional Environment of Moghra Formation

The present observations on the paleoenvironmental settings of the Early Miocene Moghra Formation have resulted in three inter-fingering environmental interpretations: it is likely that the Moghra Formation was deposited in three environments: (1) Tide-dominated Estuary (Facies Association 1), (2) Open Shelf (Facies

S. M. Hassan, *Sequence Stratigraphy of the Lower Miocene Moghra Formation in the Qattara Depression, North Western Desert, Egypt*, SpringerBriefs in Earth Sciences, DOI: 10.1007/978-3-319-00330-6_7, © The Author(s) 2013

Table 7.1 Summary of facies and facies associations

Depositional environment	Facies Association		Facies
1. Tide dominated estuary	(A) Axial	Tide-influenced fluvial channel (FA1)	Channel in-fill deposits (mixed-load fluvial-tidal facies succession (F1)
		Flat laminated sandstone (FA2)	Flat-laminated sandstone (F2)
		Outer estuary bar (FA3) (Tidal Channel and Tidal Bars)	Tabular and trough cross-stratified sandstones (F3)
			Lensoidal calcareous cross-stratified sandstone (F4)
			Sandstone with *Ophiomorpha* burrows (F5)
	(B) Marginal Estuary	Tidal Flat (FA4)	Inclined heterolithic stratification IHS (F6)
			Heterolithic and Rhythmite beds (F7)
			Mudstone beds (F8)
			Mangroves within cross-bedded sandstone (F9)
2. Open shelf		Bioturbated fossiliferous sandstone (FA5)	Sandstones with moderately to intensively bioturbated branched network *Ophiomorpha* (F10)
			Calcareous, mottled, homogenized, highly bioturbated (*Ophiomorpha*) sandstone (F11)
			Fossiliferous calcareous sandstone, highly bioturbated with *Ophiomorpha* and *Thalassinoides* (F12)
			Fossiliferous ferruginous large scale cross-bedded sandstone with hard crust (F13)
			Glauconitic trough cross-bedded sandstone (F14)
3. Tide dominated delta		Bioturbated fossiliferous carbonate (FA6) Coarsning upward (FA7)	Fossiliferous limestone (F15)
			Coarse grained Lensoidal heterolithic (F16)
			Thin laminated sand—shale intercalation (F17)
			Homogenized bioturbated sand (F18)
			Bioturbated heterolithic sandstone (interbedded mudstone and sandstone) (F19)
		Fining upward (FA8)	Cross-stratified sandstone with mudstone drapes (F20)

Association 2), and (3) Tide-dominated Delta (Facies Association 3). Facies Association 2 was broadly contemporaneous with and formed basinward of both Facies Associations 1 and 3. Facies Associations 1 and 3 would have occupied the same position (i.e. the fluvial to marine transition) at different times. The lower part of (2) in any outcropping facies succession would genetically relate to (1), whereas the upper part of (2) would be broadly co-eval with (3).

The paleoenvironmental conclusions are based partly on paleoecological interpretations. Sedimentary facies and architecture favor depositional settings related to tidal channels and tidal bars of various scales and paleogeographic positions within an estuary and its genetically related, underlying delta.

7.2 Sequence Stratigraphy

The Moghra Formation is characterized by a repetitive cyclicity of transgressive-regressive tracts (TST–HST), with TSTs typically being bounded beneath by SB and above by MFI and characterized, in relatively seaward localities, by tidal channel deposits and, landward of the tidal channel, by open marine deposits. Highstand systems tracts are bounded beneath by MFI, above by SB and are characterized by delta deposits in HST. No lowstand systems tracts or major downshift surfaces have been recognized within the succession. This distinctive TST–HST cyclicity is interpreted to be an alternation between times of large ratio of accommodation space to sediment supply (TST) and then an interval of greatly reduced A/S ratio (HST).

Eleven regional stratigraphic discontinuities surfaces have been identified in the lower Miocene Moghra Formation in the Qattara Depression of northwestern Egypt. These surfaces were traced in high-resolution multi- channel across 40 km almost all the entire area. The Moghra Formation, which is exposed in a series of south-facing escarpments, consists of ~200 m of estuarine and marginal marine sandstone, siltstone, shale, and minor limestone, arranged into a series of transgressive-regressive. The cycles, reaching thicknesses of up to 45 m each, are dominated by transgressive deposits. The lower part of each cycle consists of unconsolidated cross-stratified sand and sandstone with locally abundant vertebrate fossils, petrified logs, *Thalassinoides* and *Ophiomorpha* burrows. These deposits, which are interpreted as a backstepping estuarine channel complex, are capped by pervasively bioturbated *Ophiomorpha*-bearing sandstone beds with open marine fauna. In some cases, the marine beds pass upwards into a thin interval of coarsening upward regressive shale and siltstone, such deposits are represented tide dominated delta and commonly truncated by the next master erosion surface (sequence boundary). Provisional ages, based on a combination of biostratigraphy and strontium isotope stratigraphy, range from 21 to 17 Ma for these all eleven sequences. The occurrence of some 10–11 transgressive-regressive cycles during this 4 my interval (Table 7.2) strongly suggests that the cycles represent on average some 200–400 ky, i.e., 4th -order sequences. However, regional subsidence is required to account for most of the observed stratigraphic thickness. We believe that the preferential preservation of transgressive half cycles is due to the filling of available accommodation during higher-order sea-level rises, alternating with erosive bypass of sediment to the Mediterranean margin during sea-level falls. The poorly preserved shelf-to-deltaic half cycles represent reduced rates of relative sea-level rise (or increased rates of sediment supply) prior to each relative sea level fall.

Table 7.2 Summary table showing (1) age and duration of the different T-R sequences

Age (Ma)	Sequences	T-R (Tracts)	Shoreline Trajectories
17 16.5?	S11	Transg.	
————	S10	Transg.	
17.4 18.2	S9	Reg.	
————		Transg.	
————	S8	Reg.	
————		Transg.	
18.2?	S7	Reg.?	
————		Transg.	
————	S6	Reg.?	
————		Transg.	
————	S5	Reg.?	
————		Transg.	
————	S4	Transg.	
————	S3	Transg.	
————	S2	Reg.?	
19.6		Transg.	
21	S1	Reg.	
			S L

S = Seaward; L = Landward

7.3   The Broader Setting of the Moghra Estuaries

Thus there are two alternatives, one involving a very large, master incised valley, and the other a non-valleyed scenario for the Moghra estuarine-delta system.

Alternative Valleyed and Non-Valleyed Interpretations of Moghra Estuarine Complex

In the present study, we document a depositional system that has all the characteristics, in terms of depositional facies, of a well-developed, repeatedly-stacked, estuary-delta system. However, the Moghra system possibly lacks evidence of an obvious sub-aerial unconformity at the erosive base of the entire Moghra complex. For this reason the non-valleyed alternative is a likely and would be consistent with Dalrymple's (2006) recent modification of the definition of an estuary (Løseth et al. 2009). In addition, we prefer a non-valleyed interpretation for the Moghra succession, mainly because there is no field evidence for master valley walls to enclose the entire succession. The alternative, the subsiding tectonic depression model, is therefore more consistent with the stratigraphic evidence. Regarding the cyclic (repeated transgressive-regressive) character of the entire infill, we reject high-frequency tectonic uplifts or eustatic sea-level falls to explain the repeated transgressive half-cycles, again because there is little supportive evidence. On the other hand, with the overall subsidence context (causing overall relative sea level rise), the regressive half cycles can be simply and economically explained by climate-driven sediment supply increases. Increased sediment flux during subsidence-driven transgression is sufficient to cause a short-term regression turnaround in the stratigraphy. Nevertheless the transgressive half cycles (the estuaries) do dominate the succession, requiring only very short punctuations of increased supply to interrupt the dominant transgressive tendency.

7.4   Recommendations for Future Work

1 We already have Burdigilain ages for the study succession. We aim to identify the boundary between the lower Miocene and Oligocene. In addition, we plan to age date the fossils around this contact.
2 We have already established a sequence stratigraphy in the area, but will continue search for further robust key surfaces between the established master sequence boundaries.
3 Although the area is rich in ichnofacies like the *Ophiomorpha* and *Thalassiniodes*. There has been no thorough ichnology study of the area. Therefore, we

prepare for cooperation with some professors of ichnology to give further interpretation and more detailed examination of these trace fossils.

4 A paleobotanical study is required to throw more light on the thoroughly worm-bored petrified trees including the Palm trees. No palaeobotonical work has been done on these petrified trees and fossil mangroves until now.